Kai Uhrig

Investigation of Cytoskeletal Systems with Optical Tweezers

Kai Uhrig

Investigation of Cytoskeletal Systems with Optical Tweezers

Adhesion Forces of Actin Filaments and Malaria Parasites Measured with Optical Tweezers in Microfluidic Environments

Südwestdeutscher Verlag für Hochschulschriften

Impressum/Imprint (nur für Deutschland/ only for Germany)
Bibliografische Information der Deutschen Nationalbibliothek: Die Deutsche Nationalbibliothek verzeichnet diese Publikation in der Deutschen Nationalbibliografie; detaillierte bibliografische Daten sind im Internet über http://dnb.d-nb.de abrufbar.

Alle in diesem Buch genannten Marken und Produktnamen unterliegen warenzeichen-, marken- oder patentrechtlichem Schutz bzw. sind Warenzeichen oder eingetragene Warenzeichen der jeweiligen Inhaber. Die Wiedergabe von Marken, Produktnamen, Gebrauchsnamen, Handelsnamen, Warenbezeichnungen u.s.w. in diesem Werk berechtigt auch ohne besondere Kennzeichnung nicht zu der Annahme, dass solche Namen im Sinne der Warenzeichen- und Markenschutzgesetzgebung als frei zu betrachten wären und daher von jedermann benutzt werden dürften.

Verlag: Südwestdeutscher Verlag für Hochschulschriften Aktiengesellschaft & Co. KG
Dudweiler Landstr. 99, 66123 Saarbrücken, Deutschland
Telefon +49 681 37 20 271-1, Telefax +49 681 37 20 271-0, Email: info@svh-verlag.de
Zugl.: Heidelberg, Ruprecht-Karls-Universität, Diss., 2009

Herstellung in Deutschland:
Schaltungsdienst Lange o.H.G., Berlin
Books on Demand GmbH, Norderstedt
Reha GmbH, Saarbrücken
Amazon Distribution GmbH, Leipzig
ISBN: 978-3-8381-1046-2

Imprint (only for USA, GB)
Bibliographic information published by the Deutsche Nationalbibliothek: The Deutsche Nationalbibliothek lists this publication in the Deutsche Nationalbibliografie; detailed bibliographic data are available in the Internet at http://dnb.d-nb.de.
Any brand names and product names mentioned in this book are subject to trademark, brand or patent protection and are trademarks or registered trademarks of their respective holders. The use of brand names, product names, common names, trade names, product descriptions etc. even without a particular marking in this works is in no way to be construed to mean that such names may be regarded as unrestricted in respect of trademark and brand protection legislation and could thus be used by anyone.

Publisher:
Südwestdeutscher Verlag für Hochschulschriften Aktiengesellschaft & Co. KG
Dudweiler Landstr. 99, 66123 Saarbrücken, Germany
Phone +49 681 37 20 271-1, Fax +49 681 37 20 271-0, Email: info@svh-verlag.de

Copyright © 2009 by the author and Südwestdeutscher Verlag für Hochschulschriften Aktiengesellschaft & Co. KG and licensors
All rights reserved. Saarbrücken 2009

Printed in the U.S.A.
Printed in the U.K. by (see last page)
ISBN: 978-3-8381-1046-2

Contents

Abstract   **6**

I   Introduction   7

1   Introduction and Outline   **9**

2   Theory of Optical Tweezers   **13**
- 2.1   Principle of Optical Forces 13
- 2.2   Geometrical Optics Regime 19
- 2.3   Rayleigh Regime 23
- 2.4   External Factors on Particle Dynamics 25
- 2.5   Calibration of Optical Traps 27
  - 2.5.1   Equipartition Theorem 28
  - 2.5.2   Statistical Analysis of Position Distribution 29
  - 2.5.3   Hydrodynamic Friction and Stokes Force 29
  - 2.5.4   Power Spectra Analysis 31
- 2.6   Holographic Optical Tweezers 34
  - 2.6.1   Multiple Trap Systems 34
  - 2.6.2   Theory of Diffraction and Holography 36
  - 2.6.3   Fourier Optics 37
  - 2.6.4   Computer Generated Holograms 38

3 Actin and the Cytoskeleton — 45
3.1 Cytoskeletal Components — 45
3.1.1 Microtubules — 45
3.1.2 Intermediate Filaments — 46
3.1.3 Actin — 47
3.1.4 Biochemistry and Biology of Actin — 47
3.1.5 Actin Binding Proteins — 50
Myosin II — 50
α-Actinin — 51
3.1.6 Actin Cortex and Actin Bundles — 52

4 Malaria and Biophysics of the Malaria Parasite — 55
4.1 History and Background of Malaria — 56
4.2 Biology and Life Cycle of the Malaria Parasite — 57
4.3 Motility and Adhesion of *Plasmodium* Sporozoites — 59

II Materials and Methods — 65

5 Microscopy Setup — 67
5.1 Holographic Optical Tweezers Setup — 69
5.2 Fluorescence Microscopy Setup — 69
5.3 Brightfield Setup — 71
5.3.1 High Magnification Inverted Microscope Setup — 71
5.3.2 Low Magnification Upright Microscope Setup — 72
5.4 Setup Control — 72
5.5 Tracking Procedures — 75

6 Microfluidic Flow Cells — 77
6.1 Photolithography — 78
6.2 PDMS Casting — 80
6.3 Assembly of Flow Cells — 81
6.4 Cleaning of Flow Cells — 81
6.5 Flow Cell Design and Scheme of Usage — 83
6.6 Surface Modification of Micropillars — 84

7 Proteins and Buffer Solutions   85
    7.1 Actin Isolation and Purification 85
    7.2 Actin Storage and Dialysis 86
    7.3 Actin Polymerization and Staining 86
    7.4 α-Actinin 88
    7.5 Adhesive Microparticles 89

8 Plasmodium Sporozoite Experiments   91
    8.1 Preparation of *Plasmodium berghei* sporozoites 91
    8.2 Optical Tweezers Experiments with Sporozoites 92

III Experiments and Results   93

9 Two-Dimensional Actin Networks   95
    9.1 Experimental Procedure to Create and Probe Two-Dimensional Actin Networks 96
       9.1.1 Calibration of Trap Pattern 102
    9.2 Contractile Forces during Cross-Linking of Actin Network 104
    9.3 Conclusions on Two-Dimensional Actin Networks 107

10 Zipping Forces Between Filaments in HOT   111
    10.1 Experimental Procedure 112
    10.2 Data Analysis and Results 114
    10.3 Conclusions on Actin Snapping Experiments 118

11 Unzipping Forces Between Filaments in HOT   121
    11.1 Background for the Induced Unzipping of Semiflexible Polymers . . . 121
    11.2 Experimental Procedure 124
    11.3 Data Analysis and Results 126
    11.4 Conclusions on Actin Unzipping Experiments using HOT 128

12 Unzipping Forces Between Filaments on Pillar Substrates   131
    12.1 Experimental Procedure 132
    12.2 Data Analysis and Results 136
    12.3 Conclusions on Actin Unzipping Experiments on Pillar Substrates . . 139

13 Probing Adhesion of *Plasmodium* Sporozoites with Optical Tweezers   **141**
    13.1 Viability of Sporozoites in Optical Traps 142
    13.2 Force Calibration for Trapped Sporozoites 142
    13.3 Adhesion Experiments with *Plasmodium* Sporozoites 147
    13.4 Results and Discussion of Sporozoite Experiments 150
    13.5 Conclusions and Outlook for *Plasmodium* Experiments 150

14 Discussion and Outlook   **153**

List of Figures   **157**

Bibliography   **161**

A Publications   **189**

Abstract

Optical tweezers are a versatile tool to apply and measure forces in the piconewton range on microscopic objects that are held by optical forces in a focussed laser beam. We employed holographic optical tweezers (HOT) to create extended force sensor arrays, consisting of multiple trapped particles that were controlled and probed individually. The combination of high-speed video microscopy with fluorescence imaging allowed the visualization of labeled protein structures in parallel with the tracking of multiple trapped particles for force measurements. Using this setup, we could perform quantitative force measurements on biological samples with HOT for the first time.

To obtain reliable force measurements, calibration methods based on power spectra analysis were adapted for holographic optical tweezers. In microfluidic environments, biomimetic structures of the cellular cytoskeleton could be reconstituted between optically trapped microspheres. Flow cells and fluidic control, developed in this work allowed the exchange of solutions in the system and thus, the complete control of the chemical environment without generating forces that would affect the trapped particles.

This provided the possibility to measure dynamic processes such as the contractility of two-dimensional cross-linked actin networks in the microfluidic flow cell. A network of actin fibers between seven trapped particles was created and the forces during cross-linking were obtained.

To investigate bundling processes between filaments, a method has been established using dynamic HOT to manipulate zipper-like structures between trapped particles actively. Analysis of particle trajectories during zipping of filaments allowed the determination of binding energies between filaments.

Unbundling forces between actin filaments were measured on trapped spheres during the active process of unzipping. Additionally, this system was transferred to actin networks on PDMS micropillar substrates to improve feasibility. In combination with the optical trap, this allowed for the investigation of unbundling forces for α-actinin as well as for magnesium ions as cross-linkers.

In a different set of experiments, the adhesion process of the Malaria causing parasite *Plasmodium* was investigated. Adhesion and locomotion of the sporozoites is crucial for the infectivity of the parasite. A methodology for laser tweezer experiments with *Plasmodium* sporozoites was developed. Using optical tweezers, the formation of adhesion sites in the presence of actin disrupting drugs was probed and compared to knock-out parasite strains. We found the second step of sporozoite adhesion sequence to be significantly dependend on actin and a specific transmembrane protein named TRAP.

Kurzzusammenfassung

Optische Pinzetten stellen ein vielseitiges Werkzeug für die Erforschung und die Anwendung molekularer Kräfte dar. Dabei werden mikroskopische Partikel durch optische Kräfte im Fokus eines Lasers gefangen, die anschließend als Kraftsensoren verwendet werden können. In dieser Arbeit wurden holographische optische Pinzetten eingesetzt, um ausgedehnte Mehrpunktkraftsensoren zu generieren, die aus mehreren parallel gefangenen Polystyrolkügelchen bestanden. Durch die Kombination von Hochgeschwindigkeits-Videomikroskopie und Fluoreszenzmikroskopie war es möglich, sowohl die in der Falle gehaltenen Mikropartikel als auch Proteinstrukturen zwischen den Partikeln abzubilden und dynamische Prozesse zu verfolgen. Zur Kalibration der optischen Kräfte wurde die Analyse von Leistungsspektren auf die holographisch optischen Pinzetten angewandt. Dies erlaubte die zuverlässige Kalibrierung auch in der Gegenwart von externen Störungen, wie sie in Mikrofluidiksystemen häufig vorkommen.

Der Einsatz von Mikrofluidiksystemen ermöglichte die gezielte Generierung und Modifizierung biomimetische Proteinstrukturen. Hierfür wurde eine spezielle Mikrofluidikplattform entwickelt, die den Austausch von Medien in der Umgebung der optischen Fallen erlaubte ohne die Messungen zu beeinträchtigen. Bis zu sechs unabhängige Kanäle konnten einzeln in Pikolitergenauigkeit angesteuert werden, um so komplexe Experimente in der Flusszelle zu ermöglichen.

Exemplarisch für die Fähigkeiten des Systems wurde ein zweidimensionales Aktinnetzwerk als Modell für den Zellcortex erzeugt und die Kräfte bei der Kontraktion durch anschließende Quervernetzung der Filamente gemessen.

Zur Bestimmung der Adhäsionsenergie zweier Aktinfilamente, wurden Partikeltrajektorien während des Bündelungsvorganges analysiert. Aus den hydrodynamischen Reibungskräften ließ sich die Bündelungsenergie pro Längeneinheit des Filaments bestimmen.

Bündelungskräfte zwischen einzelnen Aktinfilamenten konnten durch aktive Krafteinwirkung mittels holographischer optischer Pinzetten und in einem kombinierten Aufbau aus optischen Pinzetten und Mikrosäulensubstraten gemessen werden.

In weiteren Experimenten wurde der molekulare Hintergrund der Adhäsionssequenz von *Plasmodium* Sporozoiten an Oberflächen untersucht. Die Gattung *Plasmodium* stellt den parasitären Erreger von Malaria, wobei Sporozoiten das motile Stadium des Parasiten darstellen, welches von Moskitos auf den Endwirt übertragen wird. Die

Interaktion mit Oberflächen im Sporozoiten Stadium spielt eine wichtige Rolle beim Infektionsprozess im Menschen. Die Rolle des Transmembranproteins TRAP und von F-Aktin für die Ausbildung von Adhäsionen wurde mit den optischen Pinzetten untersucht. Es konnte gezeigt werden, dass eine initiale Adhäsion auch in der Gegenwart von Aktin zerstörenden Substanzen sowie für Parasiten ohne TRAP möglich ist, die Bildung eines zweiten Anheftpunktes für die Sporozoiten jedoch von TRAP und Aktin abhängt.

Part I

Introduction

Chapter

1

Introduction and Outline

Cells are the basic structural unit of all life on earth. The chemical composition of cells does not significantly differ from their environment and the basic principles of chemistry and physics apply to all biological objects as for any other system on earth. But unlike lifeless matter, cells are able to reproduce, to grow, to adapt to their environment or to make a long story short, to live. Life itself however, is nothing else than an enormous number of chemical reactions and physical processes that follow basic principles of natural laws. But even the simplest cells represent physico-chemical systems that are much more complex than any man made machine on earth. Every single cell has to provide concentration gradients not only to its environment but also inside its cytoplasm. Cells have to perform polymerization processes in one place while the same type of molecule has to be depolymerized at another site in the cytosol. Molecules have to be hydrolyzed while others are condensed. Also, non-covalent interactions between molecules are continuously formed and released. Membrane compartments are internalized and externalized or fuse with intracellular structures. Proteins interact to form macromolecular assemblies and filaments. These filaments in turn are assembled to form cytoskeletal structures that span the whole cell body [1]. Upon mechanic stimuli, they rearrange this structures and adapt their complete shape to their environment [2, 3]. So, how are cells able to control this gigantic number of chemical and physical processes to produce this concerted interaction that we call life?

The key of biological reactions is their specificity. Proteins and other macromolecules react with their substrates very selectively; an interaction that is often termed a lock and key model. Moreover, cytoskeletal structures allow the cells to produce asymmetric structures, a polarity in their shape or the distribution of intracellular compartments. In order to investigate intracellular processes in detail it is therefore necessary to take a closer look at such specific interactions and on how cells model their cytoskeleton such that it serves their purposes best. However, investigating this processes *in vivo* exceeds the experimental possibilities given today very quickly. Cells consist of thousands of different proteins that possibly interact with each other [4]. To measure interactions between specific molecules is hardly possible in such an ensemble. One possible strategy is to reduce the complexity of the system by taking only parts of interest and copying cellular structures *in vitro*. This approach is called biomimetics. The idea is to probe the properties of a reduced system that is similar to the natural one in order to draw conclusions for the properties of the whole cell [5, 6].

Different biological systems have already been reconstituted *in vitro* to investigate the underlying principles. For example, complete protein expression systems have been constructed inside phospholipid vesicles to mimic protein production inside cells [7]. Motility assays consisting only of molecular motors and cytoskeletal filaments have been used to investigate the principles of intracellular transport and motility [8, 9]. Mechanics of single motors were also probed by optical tweezers experiments [10–12].
Optical tweezers are a versatile tool in biophysical studies due to their unique ability to arrange and probe objects on the microscopic scale. The momentum carried by light is exploited to trap objects in the potential well formed by a highly focused laser beam [13–15]. They are used in a wide range of applications, including microrheological studies, enzyme mechanics and cell adhesion experiments [16–18]. The introduction of dynamic holographic optical tweezers and, therefore, the possibility to create and to control almost arbitrary numbers and arrangements of optical traps extended the field of possible applications for optical trapping experiments even more [19–21].
However, to create more complex experiments, not only the arrangement of the traps has to be controlled, but also the control over the chemical environment must be established. This means, that it must be possible to exchange solutions without affecting the optical traps. Microfluidic systems emerged in the past years, providing the capability to exchange solutions in a highly controlled manner [22–25]. First approaches to implement optical tweezers into microfluidic technologies proved

already the enormous potential and versatility of this combination [26–30].

The goal of this work was, to develop tools for the creation of biomimetic cytoskeletal structures and to investigate their mechanical properties. In a first step, the system had to be reduced to the minimum number of required components. In our case, these were filamentous actin and cross-linking agents. But to mimic a biological system, its structure must be imitated too. The shape of the actin cortex in cells and, thus, its mechanical properties are characterized by its quasi two-dimensional structure that is confined by anchor points in the cell membrane [31, 32]. Therefore, particles arranged and held by holographic optical tweezers in a two-dimensional pattern were used as a template for the creation of a cortex like filament network. Cells actively change the structure of their cortical actin network by cross-linking filaments to form bundles [33, 34]. To investigate the dynamic contraction process during cross-linking, a new method of exchanging solutions in the biomimetic system had to be developed. A microfluidic system was employed for this purpose that had to overcome several challenges. The exchange of solutions had to be sufficiently fast in order to guarantee a reasonable experiment time, which was limited by bleaching and photo damage to the fluorescently labeled filaments. On the other hand, the exchange of solutions had to be arranged in such a way that hydrodynamic forces would not exceed the trapping strength of the holographic optical tweezers, limited to several piconewton. In addition, during force measurements the absence of flow had to be guaranteed entirely in order to provide the desired force resolution of the system. To investigate cross-linking forces between single filaments, the system's complexity was further reduced. Single filaments were confined between optically trapped particles and, in a modified approach, between elastomer micropillars. Subsequently, the mechanical properties of bundling and unbundling were probed with optical tweezers. Such bundling processes can be influenced by bending modes and thermal motion of the filaments [35, 36]. Thus, a setup where filaments are free to fluctuate should provide a more natural environment compared to unbinding experiments on surfaces [37, 38].

In further experiments, the adhesion process of malaria causing *Plasmodium* sporozoites was investigated using optical tweezers. Once sporozoites are transmitted during the blood meal of a female mosquito, they actively migrate through tissue to eventually find and invade a capillary to enter the blood circulation [39]. In the blood system, they passively float until they arrive in the *vena porta* of the liver. The parasite has to invade the hepatic tissue to continue its life cycle. Thus, adhesion to the endothelium of the *vena porta* is the first and most crucial precondition for its survival. Although it is believed that this step requires a multi step process, involving different adhesive proteins, so far no experimental approach has been performed to test this hypothesis. Here, we report the first measurements that prove the existence of a multi step adhesion process, which relies on the ability to distinguish different attachment sides. Moreover, the quality of the sequence is characterized by identifying molecular key factors for the second out of three steps in this process. In summary, these results provide new insights and a better understanding to tackle this bottleneck in the life cycle of *Plasmodiums*. In detail, optical tweezers were used in combination with either wild type (WT) parasites or strains lacking the protein TRAP (Thrombosponin Related Anonymous Protein, TRAP-KO), which was suggested to play a role in sporozoite adhesion earlier [40–42]. WT and TRAP-KO sporozoites were also tested for their second adhesion when lacking filamentous actin due to the influence of actin disrupting drugs such as cytochalsin D and swinholide. Here, the complexity of the system was reduced by eliminating potentially involved interactors and observing the resulting variations in the experiment.

This work is divided into three parts: In the first part, an introduction into the involved systems, namely the optical traps (chapter 2), the cytoskeleton (chapter 3) and the malaria parasite (chapter 4) is given. After describing the employed materials and methods (chapters 5, 6, 7 and 8), the developed experiments and the results obtained with these methods (chapters 9, 10, 11, 12 and 13) are presented.

Chapter

2

Theory of Optical Tweezers

2.1 Principle of Optical Forces

Already in the beginning of the 17th century the idea of light bearing a momentum emerged in science. In 1619, the German scientist Johannes Keppler (1571-1630) postulated that the fact of comets tails always pointing away from the sun was due to a radiation pressure of the sunlight [43]. And indeed, the same forces that act on interstellar particles in the light of a star, are responsible for the forces that can be exerted on a particle held in the laboratory by an optical tweezer.
In classical mechanics, a force F can only be exerted on an object, when a momentum p is transferred:

$$F = \frac{dp}{dt}. \tag{2.1}$$

This law is universally applicable and can therefore also be used to describe interactions between light and matter.
Following Albert Einsteins (1879-1955) interpretation of the photoelectric effect, one can attribute to every photon a certain energy

$$E_{Ph} = h\nu \tag{2.2}$$

and a momentum

$$p_{Ph} = h/\lambda. \tag{2.3}$$

Figure 2.1: Image of the comet Lulin (C/2007 N3) that approaches the earth in spring 2009. The tail of the comet, consisting of dust particles pushed by the radiation pressure of the sun points to the right. The part of the tail that seems to point to the other direction, the so called anti-tail is an optical illusion caused by the viewing geometry and typically occurs when Earth crosses the plane of a comet's orbit. Image by Paolo Candy (Cimini Astronomical Observatory).

h is the PlanckÕs constant $6.626 \cdot 10^{-34}$ Js (Max Planck, 1858-1947) and ν and λ are the frequency and the wavelength of the light[1]. Photons that hit a surface can transfer momentum and exert thus a force on the surface, when reflected or absorbed. To get an expression for the observed radiation pressure P_r on an illuminated surface A it makes sense to introduce a system equivalent to a gas pressure. Pressure is defined as a force F exerted on a surface A:

$$P_r = \frac{F}{A} = \frac{dp}{dt} \cdot \frac{1}{A}. \qquad (2.4)$$

For photons, the reflectivity of the illuminated surface determines the transferred momentum. An absorbed photon transfers a momentum of p, while a reflected one transfers $2p$, since after reflection it carries an inverted momentum $-\vec{p}$. Hence, for an illuminated surface, the transferred momentum can be written as

$$\Delta p = \gamma p, \qquad (2.5)$$

---

[1] The expression for the momentum of the photon can be derived, when considering the relativistic energy $E^2 = (cp)^2 + (mc^2)^2$ for a particle that has no rest mass m. c is here the speed of light ($3 \cdot 10^8$ m/s).

2.1 Principle of Optical Forces

where γ is the coefficient of reflectivity, which equals 1 for totally absorbing surfaces and 2 for totally reflective ones. Photons travel at the speed of light, $c = 3 \cdot 10^8 \, \text{m/s}$.

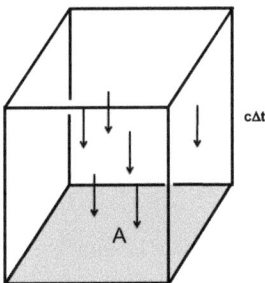

Figure 2.2: Photons in a volume $V = A \cdot c\Delta t$ will hit the surface A in the time interval Δt.

Thus, all photons that are in the volume $V = A \cdot c\Delta t$ will hit the surface A in the time interval Δt (see figure 2.2). This number of photons will be defined as N. The radiation pressure exerted by all this photons is thus given by

$$P_r = \frac{F}{A} = N\frac{\gamma p}{\Delta t} \cdot \frac{1}{A}. \tag{2.6}$$

Since $V = A \cdot c\Delta t$ it is possible to replace $A \cdot \Delta t$ with V/c. Equation 2.6 reads thus as:

$$P_r = \gamma c p n, \tag{2.7}$$

where n is the photon density:

$$n = \frac{N}{V}. \tag{2.8}$$

Now, one can define the intensity I of a light beam by the transferred power P on an area A:

$$I = \frac{P}{A} = \frac{N E_{Ph}}{\Delta t} \cdot \frac{1}{A}. \tag{2.9}$$

Using again the relation $V = A \cdot c\Delta t$ and equation 2.8 we derive an expression for the radiation pressure of a light beam of intensity I that illuminates a surface with the coefficient of reflectivity γ:

$$P_r = \gamma \frac{I}{c}. \tag{2.10}$$

Thus, an intuitive expression for the light pressure results, which is proportional to the incident intensity. This is the basis for light interacting with matter and exerting measurable forces on illuminated objects.

Similarly, when considering light as an electromagnetic wave it is possible to find an expression for the radiation pressure on an illuminated surface. The energy flux of an electromagnetic field is given by the Cartesian product of the vectors of the electrical field $\vec{E}$ and the magnetic field $\vec{B}$,

$$\vec{S} = \frac{1}{\mu_0} \vec{E} \times \vec{B}, \qquad (2.11)$$

where μ_0 is the magnetic field constant ($4\pi \cdot 10^{-7}$ V s/A m). $\vec{S}$ is the Poynting vector, named after the British physicist Sir John Poynting (1852-1914). In an electromagnetic wave, where $\vec{E}$ and $\vec{B}$ are always orthogonal, $\vec{S}$ defines the power and the direction of wave propagation. The intensity or irradiance I of an electromagnetic wave is defined by the average transferred power $\langle P \rangle$ on an area A. It can be calculated by the time-averaged magnitude of the Poynting vector $\vec{S}$ through the relationship:

$$I = \frac{\langle P \rangle}{A} = \langle |\vec{S}| \rangle_t. \qquad (2.12)$$

Analogously to equation 2.10, the radiation pressure of an electromagnetic wave can therefore be calculated as:

$$P_r = \gamma \frac{I}{c} = \gamma \frac{\langle |\vec{S}| \rangle_t}{c}. \qquad (2.13)$$

For a light source of 10 W that illuminates an area of 1 mm^2, a radiation pressure of $3.3 \cdot 10^{-2}$ Pa can be calculated. This value corresponds to 33 nN, a force that can be measured only with very accurate devices.

The lack of intense and highly collimated light sources made it impossible to use this effect for long time. Figure 2.3 shows a picture of a so called light mill; in the 19th century, these toy machines were believed to be driven by light pressure. The Russian physicist Pyotr Nikolaevich Lebedev (1866-1912) was the first man to measure the pressure of light on a solid body in 1899 in a similar device nowadays known nowadays as Nichols radiometer [46]. But only the development of laser light sources in 1960 allowed to realize optical forces that could affect microscopic objects in an observable way. Arthur Ashkin (* 1922) investigated the effects of optical scattering and gradient forces on micrometer sized particles at the AT&T

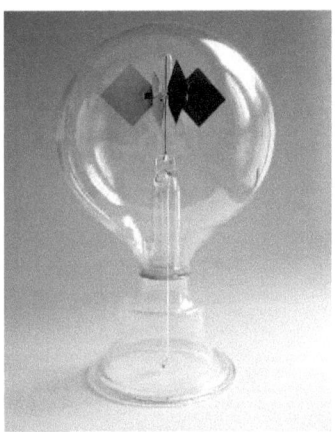

Figure 2.3: Image of a light mill. Such a Crookes (Sir William Crookes, 1832-1919) radiometer, or light mill, spins, when illuminated by a sufficiently strong light source. A light mill consists of a partially evacuated glass bulb containing a set of vanes, which are mounted on a low friction spindle. For a long time it was believed that this effect is due to the radiation pressure of the used light source. Actually, it originates from temperature differences on the different sides of the vanes and therefore induced local pressure differences [44, 45].

Bell Laboratories in 1970. He could show that the radiation force of a laser beam was sufficient to move and accelerate latex spheres, suspended in water [13]. Using two oppositely arranged laser beams he created optical potential wells, in which he could also stably trap the particles. In later experiments he employed the laser to compensate for gravitational forces of the particles and was able to construct levitation traps this way [47]. The schematic drawing of the setups used by Ashkin in his first experiments is shown in figure 2.4. The actual US secretary of Energy, Steve Chu (*1948) who worked together with Ashkin at the Bell Laboratories later developed atomic traps and cooling methods based on this work and the theoretical predictions of Ashkin, for which he was rewarded the Noble Prize in Physics in 1997 [48–51]. Also, the development of the Bose-Einstein condensate that led to another Noble Prize in Physics for Eric Cornell, Carl Wieman and Wolfgang Ketterle in 2001 was strongly influenced by this discovery [52].

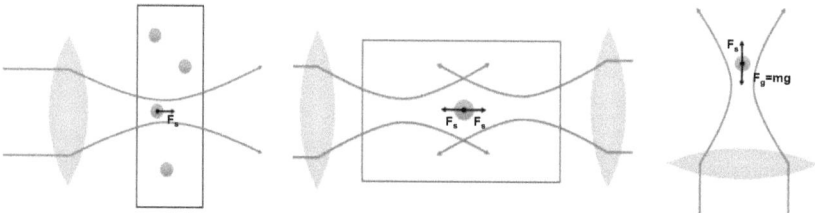

Figure 2.4: The figure shows a schematic representation of the first trapping setups realized by Arthur Ashkin. Left: Setup to accelerate particles in suspension by the scattering force $\vec{F}_s$. Middle: Trapping of particles by scattering forces of counter-propagating laser beams. Right. Optical levitation trap. The particle is held by the scattering force against its gravitational force $m\vec{g}$. These setups were first presented in [13] and [47].

Already in 1978, Ashkin proposed a different mechanism to trap objects with a highly focused laser beam, which he could finally realize in 1986: the optical tweezers [14]. Using this new technique he was able to trap and manipulate single cells, bacteria, viruses and even objects inside cells [53–55]. To understand how objects are trapped in a focused laser beam, the simple explanation by radiation pressure will no longer suffice. Instead, one has to consider the fundamental processes occurring when light passes a transparent dielectric object. To correctly account for the interaction of light and matter, the object's size has to be considered, too. For objects that are bigger than the wavelength of the incident light, geometrical optics can be applied. For smaller objects however, electrodynamic interactions between light and matter cannot be neglected. In the following chapters the origin of trapping forces for both, objects being smaller and bigger than the wavelength of the incident beam will be discussed.

2.2 Geometrical Optics Regime

For objects that are much bigger than the wavelength of the light ($r \geq 10\lambda$) simple geometrical optics, according to Snells law (Willebrord van Roijen Snell, 1580-1626) may be used. Snell's law (also known as Descartes' law or the law of refraction) is a formula used to describe the relationship between the angles of incidence and refraction, when referring to light or other waves, passing through a boundary between two different isotropic media of refractive index n_x. SnellÕs law can be written as:

$$n_1 \sin \alpha_1 = n_2 \sin \alpha_2, \qquad (2.14)$$

where α_1 and α_2 are the angles of the incident and refracted rays to the normal line of the media interface, respectively. Propagation of a light beam passing a bead can be described by this law. To illustrate this, we want to look now at a beam of power P that hits a dielectric sphere under the angle φ, being refracted at an angle ϑ, as can be seen in figure 2.5 [56]. The net force acting on the sphere can be calculated by the sum of the transferred momentum of the primary reflected ray PR and by the infinite set of subsequently refracted secondary beams of intensity $PT^2, PT^2R, PT^2R^2, \ldots, PT^2R^n$. The quantities R and T are the Fresnel coefficients of reflection and transmission of the surface at the respective angle. One can show that this sum converges and that it is possible to split up the resulting force acting on the center of the bead into gradient forces F_g along the intensity gradient of the light and scattering forces F_s into the direction of propagation of the beam [57].

$$F_s = \frac{n_M P}{c} \left\{ 1 + R \cos(2\varphi) - \frac{T^2[\cos(2\varphi - 2\vartheta) + R \cos(2\varphi)]}{1 + R^2 + 2R\cos(2\vartheta)} \right\} \qquad (2.15a)$$

$$F_g = \frac{n_M P}{c} \left\{ R \sin(2\varphi) - \frac{T^2[\sin(2\varphi - 2\vartheta) + R \sin(2\varphi)]}{1 + R^2 + 2R\cos(2\vartheta)} \right\} \qquad (2.15b)$$

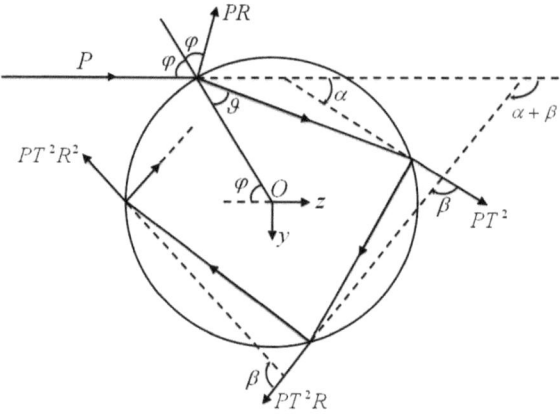

Figure 2.5: Geometry for calculating the force due to the diffraction of a single incident ray of power P by a dielectric sphere, showing the reflected ray PR and a set of refracted rays PT^2R^n (Figure adapted from [56]).

In figure 2.5, the different beams refracted by a sphere are shown schematically. The z-axis in the sketch is defined by the direction of propagation of the beam, the denotation of angles and vectors is according to equations 2.15a and 2.15b. Resulting forces depend on the polarization of the light, since R and T are different for rays polarized perpendicular or parallel to the plane of incidence [58, 59]. A typical laser beam has a defined beam profile, usually of a Gaussian shape. To get the overall force on the particle, one has to sum up the calculation shown in equation 2.15a and 2.15b for all subsidiary beams of the laser profile.

In the following, we want to give an intuitive description of the origin of optical trapping forces in a focused laser beam according to this theory. It is applicable for objects that are larger than the wavelength λ. Figure 2.6 shows the basic principle of optical trapping. If an object with a refractive index n_x differing from that of the surrounding medium is illuminated, then the light will be refracted. The refraction of the rays by the object changes the momentum of the photons according to the change in the direction from the input to the output rays. Due to the momentum conservation theorem, an equal, but opposite momentum must be transferred to

2.2 Geometrical Optics Regime

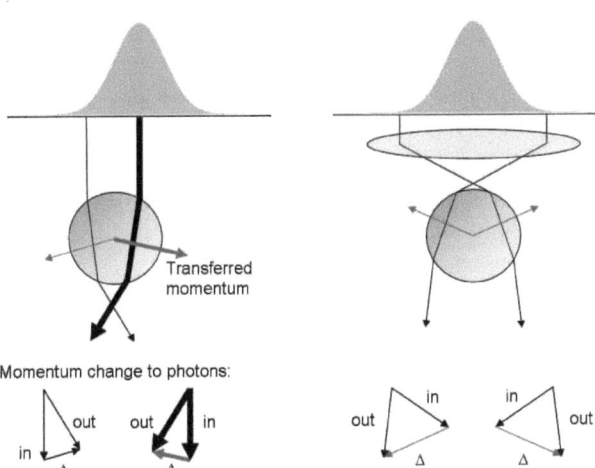

Figure 2.6: Change in the momentum of an incident laser beam of Gaussian shape on a dielectric sphere.

the object. Two cases for a spherical object are illustrated in figure 2.6: On the left, there is a parallel beam with a Gaussian profile and a bead with higher index of refraction (n_p) than that of the surrounding medium (n_m). The bead is shifted slightly to the left of the beams center. Photons hitting the bead on the left side are deflected to the right; the ones on the right side are deflected to the left, according to the law of refraction. Due to the Gaussian profile of the incoming beam, more photons are present in the center of the beam, on the right side of the bead, than on the left side, which is depicted by the thicker arrow in the sketch. Therefore, the resulting momentum transferred to the bead (when summing up the grey arrows) will point to the optical axis of the beam and downstream of the photon flux.

To achieve stable 3D trapping, a focussed light beam is necessary. This situation is shown in the right part of figure 2.6. The photons of a tightly focussed beam will gain momentum in the downstream direction of the beam when refracted by an object, which is placed behind the focus of the beam. This results in a net force on the bead towards the focus of the beam. The radial momentum components will cancel each other out if the bead is positioned in the beams axis. Otherwise, a force component towards the beams central axis, according to the left drawing will result. Combining the two situations, one can understand, that in a stable optical trap the forces on the object will always point to the place of the highest light intensity, i.e. the focus. Scattering forces on the object, which are not depicted here, will shift this stable trapping point a little bit ($< r$) downstream of the focus. Since the momentum of a single photon is very small, very high light intensities and gradients are required. Therefore, microscope objectives that are the strongest lenses available and high intensity lasers are employed in optical trapping.

Using the presented theory it is possible to calculate that in a normal laser trap, the spring constants in z direction are always weaker than in x and y, perpendicular to the direction of propagation [59, 60]. Furthermore one can see, that light rays hitting the particle under high angles of incidence, i.e. the peripheral rays, contribute much stronger to the trapping forces than the central rays. Therefore, one can improve the performance of an optical trap by overfilling the aperture of the objective in order to have higher intensity in the beams periphery. For the same reasons it would be advantageous to use the TEM_{01}^* mode of a laser instead of the "normal" TEM_{00} mode, since its "doughnought" like shape provides high intensity at the rim of the beam. However, due to its easier availability, the TEM_{00} mode is still the choice in most laboratories. Figure 2.7 shows how the available gradient

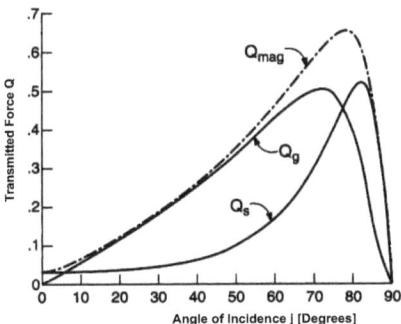

Figure 2.7: Gradient forces, scattering forces and magnitude of the total force Q_{mag} for a single ray hitting a dielectric sphere as a function of incidence angle (Figure from [56]).

forces depend on the angle of incidence, which is the reason for the use of objectives with a high numerical aperture (NA) that can provide maximum angles of incidence in optical trapping[2].

2.3 Rayleigh Regime

For particles much smaller than the wavelength of the incident light it is possible to describe the interaction between light and particle using electromagnetic theories named Rayleigh theory, after John William Strutt, 3rd Baron Rayleigh (1842-1919). Here, the particle is considered as a point-shaped dipole interacting with the electric field of the light, $\vec{E}(r,t)$ [14, 61]. The dipole moment for a spherical particle of radius r in a medium of refractive index n_m is given by:

$$\vec{p}(r,t) = 4\pi\epsilon_0 n_m^2 r^3 \left(\frac{n_{rel}^2 - 1}{n_{rel}^2 + 2} \right) \vec{E}(r,t). \tag{2.16}$$

---
[2]The numerical aperture of a microscope objective is a measure of its ability to gather light and resolve fine specimen detail at a fixed object distance. An interactive tutorial for the relation between the opening angle of an objective and the numerical aperture can be found at: http://www.microscopyu.com/tutorials/java/objectives/nuaperture/

Here ϵ_0 is the vacuum permittivity ($8.854 \cdot 10^{-12} \, C/Vm$) and $n_{rel} = \frac{n_p}{n_m}$ is the relative refractive index of the particle in the medium. Scattering forces are explained in this theory by the effect of absorption and reemission of light by the dipole; as the electric field oscillates harmonically in time, the induced point dipole follows synchronously the electric field and so, the particle acts as an oscillating electric dipole that radiates secondary or scattered waves in all directions. This scattering changes both magnitude and direction of the energy flux of the electromagnetic wave. The corresponding momentum transfer also occurs and the scattering force associating with these changes is exerted on the particle. The scattering force writes as

$$\vec{F}_s = \sigma \frac{n_m}{c} I(\vec{r}) \vec{e}_z, \tag{2.17}$$

where $\vec{e}_z$ is the unit vector in the beam propagation direction and σ is a specific scattering cross section, which represents a radius of interaction between the particle and light. For a small spherical particle that scatters the light isotropically, this cross section is given by:

$$\sigma = \frac{8}{3}\pi(kr)^4 r^2 \left(\frac{n_{rel}^2 - 1}{n_{rel}^2 + 2}\right)^2, \tag{2.18}$$

where k is the angular wave number of the light. Since emission of light is distributed over the whole three-dimensional solid angle whilst absorption is directed via the vector of propagation of the light, a net scattering force in this direction results.

The gradient force F_g can be explained by a Lorentz force (Hendrik Antoon Lorentz, 1853-1928), acting on the dipole induced by the electromagnetic field [62]. The gradient force can be calculated as:

$$\begin{aligned}\vec{F}_g(\vec{r},t) &= [\vec{p}(\vec{r},t) \cdot \nabla] \vec{E}(\vec{r},t) \\ &= 4\pi\epsilon_0 n_m^2 r^3 \left(\frac{n_{rel}^2 - 1}{n_{rel}^2 + 2}\right) \frac{1}{2} \nabla \vec{E}^2(\vec{r},t).\end{aligned} \tag{2.19}$$

The gradient force, experienced by the particle in a steady state is the time-averaged version of equation 2.19 and is given by

$$\vec{F}_g(\vec{r}) = \left\langle \vec{F}_g(\vec{r},t) \right\rangle_t. \tag{2.20}$$

Using the relation

$$\left\langle \vec{E}(\vec{r},t) \right\rangle_t = \frac{1}{2}\left|\vec{E}(\vec{r})\right|^2 = \frac{I(\vec{r})}{\epsilon_0 n_m c}, \tag{2.21}$$

one retrieves the solution

$$\vec{F}_g(\vec{r}) = 2\pi r^3 \frac{n_m}{c} \left(\frac{n_{rel}^2 - 1}{n_{rel}^2 + 2} \right) \nabla I(\vec{r}). \tag{2.22}$$

Increasing the NA decreases the focal spot size and therefore increases the intensity gradient. Hence, in the Rayleigh regime, trapping forces in all directions increase with higher NA.

In contrast to the scattering force of equation 2.17, the gradient force consists of three rectangular components, which act as restoring forces directed towards the focus of the beam-waist center in the case of $n_{rel} > 1$. These equations are an approximation for only weakly focused beams. They neglect nonuniformity of the electric field over an extent of the particle as it occurs in tight foci, where all associated rectangular components of the electric and magnetic fields can not be neglected anymore [63, 64]. The here presented deductions of optical forces give a basic view of the topic, but in literature numerous studies exist, providing exact calculations for forces that act on dielectric particles in optical traps [60, 65, 66].

In real systems, optical forces are commonly described by the relationship:

$$F = Q \frac{n_m P}{c}, \tag{2.23}$$

where Q is the dimensionless efficiency of the system. It depends on the angle of incidence, the wavelength, the polarization and the mode of the light as well as on the index of refraction and the form of the trapped particle. Q represents the fraction of power of the incident light utilized to exert force. For real trapping systems, Q typically has values between 0.03 and 0.1.

2.4 External Factors on Particle Dynamics

A particle in suspension, trapped by optical tweezers is always influenced by external forces that define his motion and behavior inside the trap. This factors can be estimated and calculated using approximations about the physical environment of the particle.

In close proximity to the optical traps the particle will feel a restoring force towards an equilibrium position, which is the trap center. This force is proportional to the displacement x and the stiffness of the optical trap, κ. It can be written as

$$F = -\kappa x \qquad (2.24)$$

equivalent to the restoring force of a spring, according to HookeÕs Law (Sir Robert Hooke, 1637-1703). This linear force-displacement relationship is valid up to a displacement of about half the particles' radius [67]. To describe the dynamics of a particle inside such a force field, one can set up an equation of motion:

$$m\frac{\partial^2 x}{\partial t^2} + \gamma \frac{\partial x}{\partial t} = -\kappa x \qquad (2.25)$$

which accounts for inertial forces, friction forces due to the motion of the particle in the medium with the friction coefficient γ and the restoring force of the trap.

At temperatures above zero, every particle is subject to permanent collisions with surrounding molecules, but unlike on macroscopic scales, this interaction cannot be neglected for microscopic systems. Therefore, microscopic objects are in continuous motion. This is the so-called Brownian motion after the Scottish botanist Robert Brown (1773-1858) who was the first to observe this thermal motion in 1827, while examining pollen grains in water under a microscope. Direction and magnitude of this motion are purely statistically distributed. According to the French physicist Paul Langevin (1872-1946) one can describe Brownian motion in a potential using a stochastic differential equation

$$m\frac{\partial^2 x}{\partial t^2} + \gamma \frac{\partial x}{\partial t} + \kappa x = F_{th}. \qquad (2.26)$$

Here, F_{th} stands for the statistical thermal force that depicts the noise term in this equation. It should follow a Gaussian distribution with an evanescent mean. Furthermore it is uncorrelated for time scales that are significantly longer than the average time between collisions (Gaussian white noise). The friction coefficient γ can be calculated for a spherical particle in a fluid medium of viscosity η according to Stokes law (Sir George Gabriel Stokes, 1819-1903):

$$\gamma = 6\pi\eta r. \qquad (2.27)$$

In an overdamped system the relation:

$$\gamma^2 > 4m\kappa \tag{2.28}$$

must be fulfilled, so inertia forces may be neglected. For a typical polystyrene particle with a radius of 1 µm and a viscosity of $\eta = 1$ mPa s, γ^2 is in the order of $1 \cdot 10^{-16}$ N kg/m. $4mk$ for a typical spring constant of 50 pN/µm and a density of polystyrene $\rho = 1.05$ g/cm^3 has a value of $1 \cdot 10^{-19}$ Nkg/m, which is three orders of magnitude smaller than γ^2. Hence, one has a strongly overdamped system where neither inertia forces nor gravitational forces play a significant role. Accordingly, the Langevin equation simplifies to

$$\gamma \frac{\partial x}{\partial t} + \kappa x = F_{th}. \tag{2.29}$$

Based on this equation one can describe the observed motion of particles inside the optical traps. It can be used to determine the spring constant κ of the trap experimentally as will be shown in the subsequent chapters.

2.5 Calibration of Optical Traps

Theoretical descriptions of optical forces are only exact in the case of the trapped particle being either much smaller or larger than the wavelength of the incident light. In most experimental setups, however, the wavelength of the laser as well as the radius of the trapped particles both are in the range of several hundred nm up to a few micrometers. This is mostly due to practical reasons like the technical availability of lasers, maximized trapping efficiency, transmission of optical elements and low absorbance of biological samples as well as handling and availability of micro particles. Therefore, it is necessary to calibrate the optical traps individually for each experiment in order to obtain reliable values for forces exerted or measured with the optical traps since exact theoretical descriptions in this regime are still lacking.

There are several different techniques used to calibrate optical traps. Svoboda and Block give an overview about the individual requirements [68]. In the following section, we will discuss some methods to calibrate the stiffness of optical traps, especially the ones used within the this work.

All of the presented methods depend on a way to determine the position of the observed particle over time. This can be done for example by video microscopy with a CCD camera. Hence, the trapped particles center must be tracked using proper algorithms. The obtained data must be transferred from pixel values to metric values as can be done using calibration grids for microscopy. It is worth to mention, that the center of a particle in the microscope can be determined with much higher accuracy than the optical resolution of $\lambda/2 \approx 200\,\text{nm}$. Since a typical particle consists in high resolution microscopy of a multitude of pixels, its center can be fitted by a Gaussian profile with an accuracy of few nanometers.

Other technologies include the determination of the particles position using quadrant photo diodes or polarization interferometric methods. These techniques have the advantage of a high dynamic range compared to standard video microscopy methods. But they lack the ability to track multiple particles simultaneously. For this reason, in our setup camera based calibration methods were chosen.

2.5.1 Equipartition Theorem

The equipartition theorem of classical statistical mechanics states that the mean value of each independent quadratic term in the energy of an object equals $\frac{1}{2}k_BT$. This means every degree of freedom of a particle at temperature T contributes with $\frac{1}{2}k_BT$ to the total energy of the system. Here, k_B is the Boltzmann's constant $(1.381 \cdot 10^{-23}\,\text{J/K})$, named after Ludwig Boltzmann (1844-1906).

To describe the motion of a particle in proximity of an optical trap, we will focus on the translational motion in the three dimensions of space. The mean potential energy of the particle in the harmonic potential of the trap can be described according to Equation 2.24:

$$E_{pot} = \frac{1}{2}\kappa\langle x^2\rangle, \qquad (2.30)$$

where $\langle x^2 \rangle$ is the average displacement from the minimum of the potential. To translocate the particle from the traps center energy must be transformed into potential energy of the trap. Therefore, it is possible to set $E_{pot} = \frac{1}{2}k_BT$. We can write now:

$$\frac{1}{2}k_BT = \frac{1}{2}\kappa\langle x^2\rangle \quad \text{and accordingly} \quad \kappa = \frac{k_BT}{\langle x^2\rangle}. \qquad (2.31)$$

It is thus possible to determine the spring constant κ of the optical trap by measuring the position data of the trapped particle. Therefore, one has to record a high number data points in order to obtain an exact mean of the traps center and the mean quadratic deviation of the particles position, which are used to calculate the traps stiffness. Possible errors of this method could arise from drifts or vibrations in the system. This becomes especially important when the trap has to be calibrated at higher spring constants, because the value for the experimental deviation decreases at higher laser powers, while the error due to external vibrations stays constant. So, systematically lower spring constants will be measured. Furthermore, the temperature inside the microscopic chamber could deviate from the ambient temperature due to absorption of laser light or the microscopes illumination, which could lead to a misinterpretation of the measured data [69].

2.5.2 Statistical Analysis of Position Distribution

In thermal equilibrium the probability $P(x)$ to find a particle in a harmonic potential $U(x)$ at a position x can be described according to the Boltzmann theory as:

$$P(x) \propto \exp\left(-\frac{U(x)}{k_B T}\right). \tag{2.32}$$

Since the harmonic potential of the trap is given as

$$U(x) = \frac{1}{2}\kappa x^2, \tag{2.33}$$

it is possible to fit a Gaussian distribution to the position data of the particle inside the force field of the trap and obtain the relevant spring constant κ from this distribution. Here, it is important to mention that in order to get a reliable fit we need position data also for higher deviations from the trap center. Therefore, large data sets and short exposure times are needed to avoid artifacts. On the other hand, the shape of the measured distribution gives information about the real shape of the traps potential, which can deviate significantly from a pure harmonic form.

2.5.3 Hydrodynamic Friction and Stokes Force

A particle that is suspended in a viscous medium will feel frictional forces when a relative motion between the particle and the medium is present. This force can be employed to calibrate the stiffness of the optical trap. Therefore, either the particle is moved through the medium or a flow has to be applied to a stationary trapped particle. This forces are proportional to the applied relative velocity v and to the friction coefficient γ (see 2.28) of the particle in the medium,

$$F = \gamma v. \qquad (2.34)$$

This force will push the particle out of the traps center until the restoring force of the trap will come to an equilibrium:

$$6\pi\eta r v = \kappa x \qquad (2.35)$$

Thus, the spring constant κ can be derived as:

$$\kappa = \frac{6\pi\eta r v}{x} \qquad (2.36)$$

To measure spring constants using this method, it is suitable to keep the trap constant and move the surrounding medium at a constant velocity. This can be done easily, using a computer controlled piezo stage. The applied velocity must be chosen such, that the viscous drag is higher than the noise due to brownian motion to get a clear signal. At the same time, the velocity must not be to high in order not to leave the linear regime of the restoring forces. Further measurement errors can arise due to displacements of the particle out of the focal plane at higher transversal forces [67].

Moreover, it is important to keep in mind, that flow fields next to surfaces must be corrected according to Faxens Law for boundary effects. They account for the fact, that the particle feels a higher friction next to a surface. Therefore, a corrected friction coefficient γ_{Fax} must be considered:

$$\gamma_{Fax} = \frac{\gamma}{1 - \frac{9}{16}(\frac{r}{h}) + \frac{1}{8}(\frac{r}{h})^3 - \frac{45}{256}(\frac{r}{h})^4 - \frac{1}{16},(\frac{r}{h})^5} \qquad (2.37)$$

r stands here for the particle radius and h for the distance from the surface. For short distances this correction can exceed 100% of the original friction coefficient. Especially in microfluidic chambers and when high NA immersion objectives with short working distances are employed this correction must be accounted for.

Finally, one has to consider, that the viscosity of water is strongly dependent on the

2.5 Calibration of Optical Traps

temperature, it can be written as [69, 70]:

$$\lg \eta_{H_2O} = \frac{1.3272\,(293.15 - T/K) - 0.0010053\,(T/K - 293.15)^2}{T/K - 168.15} - 2.99, \quad (2.38)$$

where T is here the absolute temperature in Kelvin. The viscosity varies between 20 and 37°C , which is the for biological relevant range by more than 35%.

Other variations of this technique measure the maximum velocity at which the particle escapes the trap [55, 71]. Here, the velocity is increased linearly until the particle escapes the trap. The method is applied especially in the case of biological specimen with irregular shape whose center and therefore exact position relative to the trap are difficult to determine. This way, it is possible to determine the maximum trapping strength instead of the spring constant, but the principle is comparable and the mentioned restrictions are the same.

2.5.4 Power Spectra Analysis

The thermal motion of a microscopic particle inside the harmonic force field of an optical trap can be described exactly by an equation of motion

$$\gamma \frac{\partial x}{\partial t} + \kappa x = F(t). \quad (2.39)$$

If $F(t)$ is a purely thermal force, induced by collisions with solvent molecules in the surrounding medium, then it is possible to describe the spectral density of the dynamics of the particles by a Lorentzian power spectrum of Brownian motion in a parabolic potential well [72]:

$$S(f) = \frac{k_B T}{\pi^2 \gamma (f^2 + f_c^2)}. \quad (2.40)$$

The critical frequency or "corner frequency" f_c writes as:

$$f_c = \frac{\kappa}{2\pi\gamma}. \quad (2.41)$$

This gives an elegant possibility to determine the spring constant by fitting the experimental spectrum to this equations, without the need for doing a spatial calibration of the detector. A typical power spectrum for the motion of a confined particle is shown in figure 2.8.

The spectrum can be divided into two regimes. For frequencies below the corner frequency ($f \ll f_c$) the spectral density $S(f)$ is dominated by the harmonic potential of the trap and has therefore a nearly constant value. For higher frequencies above the corner frequency ($f \gg f_c$), or very short time scales, one observes the free diffusion of the particle inside the medium, equivalent to $\kappa = 0$. In this regime, $S(f)$ follows a function $\propto 1/f^2$. Einstein proposed an expression that connects the

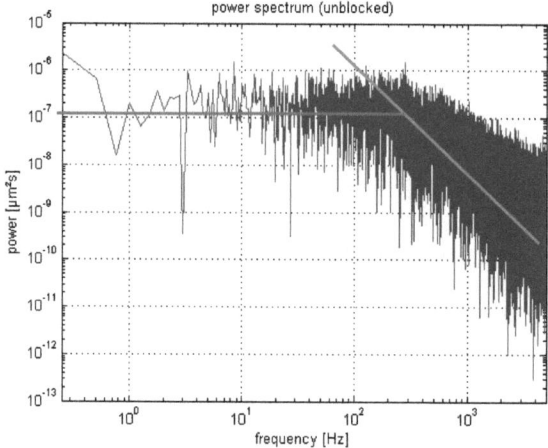

Figure 2.8: A typical power spectrum for a trapped particle. The two regimes below and above the corner frequency are marked by green and red lines.

diffusion constant D directly to the friction coefficient γ:

$$D = \frac{k_B T}{\gamma}. \tag{2.42}$$

Using this relation the spectral density can be written as:

$$S(f) = \frac{D}{\pi^2 (f^2 + f_c^2)} \tag{2.43}$$

2.5 Calibration of Optical Traps

which allows to get this experimental values directly from the obtained spectra. For particles where the friction coefficient γ is known the reliability of the measurement can hence be determined.

The power spectrum is dependent on the experimental temperature directly according to Eq.: 2.40 and indirectly via the viscosity η, which is included in γ. So variations in T can influence the measurements severely.

Typical corner frequencies can reach values up to 1000 Hz. To measure such frequencies unambiguously, the sampling frequency must be at least twice as high, according to the Nyquist Relation:

$$f_{Nyq} = \frac{1}{2} f_{sampl} \tag{2.44}$$

Typical video microscopy setups reach frame rates below 500 Hz and are therefore not suitable for this method. Methods that use photodiodes for detection overcome this restriction due to much higher frame rates. However, these methods are limited by the impossibility to track more than two particles at once. Therefore, high-speed video microscopy calibration methods at frame rates of up to several thousand Hertz have emerged in the last years [73]. In this work, we present for the first time the simultaneous calibration of extended trapping arrays with the power spectrum method.

An important advantage of the power spectrum method compared to the equipartition theorem is the fact, that most external error sources, like system vibration and drifts take place in the low frequency regime. Accordingly, it is possible to exclude them by setting the fitting ranges to appropriate values that still include the relevant corner frequencies (f_c) but exclude the noise frequencies that can be seen as characteristic peaks in the elsewhere constant frequency range below f_c.

A drawback of this method is the very high number of data points, required to get a sufficiently high frequency resolution δf. The resolution of the Fourier Transform is determined by measurement time T_m, which can be derived from the number of data points N and the frame rate δt of the image acquisition:

$$\delta f = \frac{1}{T_m} = \frac{1}{N \delta t} \tag{2.45}$$

To get a resolution of 0.1 Hz at a frame rate of 2000 Hz ($\delta t = 0.5$ ms), already 20000 frames have to be analyzed. With this frame rate power spectra up to 1000 Hz can be obtained (see Eq.: 2.44). To cover higher frequencies, higher frame rates and therefore even bigger data sets are required.

2.6 Holographic Optical Tweezers

2.6.1 Multiple Trap Systems

Optical Tweezers have become an important tool in many scientific experiments, namely force applications and force measurements in numerous biological experiments. But as more ideas to investigate more sophisticated systems in biology and colloid physics came up, the need for more complicated laser trapping setups grew. These setups should provide multiple traps that could be independently addressable and detectable. The simplest solution to create a multi trap system is to use different lasers that are independently coupled into the objective. By placing adaptable optical elements like movable mirrors or lenses into each of the different light paths the respective laser trap can individually be controlled.

To create two traps that can also be simultaneously detected, orthogonal polarized laser beams were used in a setup resembling a Mach-Zehnder interferometer (Ludwig Mach, 1868-1951; Ludwig Zehnder, 1854-1949). Here, one laser was split up using polarizing beam splitters forming two traps in the focal plane that could be tracked separately due to their polarization [74, 75].

But this technologies become very impracticable when higher numbers of optical traps are required. Another possibility to create multiple traps is the use of so called time-shared traps. Hereby, a laser is scanned cyclically between multiple focus points at such a repetition rate that trapped particles do not diffuse out of the trap in the time the laser needs to scan through the other trapping positions [64, 76, 77]. This method requires very high repetition rates, therefore acousto-optical devices are employed reaching frame rates in the kilohertz regime or even above. Time shared optical traps are the method of choice for adjustable and controllable two-dimensional trap systems. The position of each trap can be easily detected and adjusted in this setup. Trapping strength can be controlled individually by defining the time, the laser remains in each trapping point. Restrictions of this technique are that it is only possible to create patterns in the focal plane and that due to the serial nature of the traps, it is problematic to measure forces

2.6 Holographic Optical Tweezers

transmitted between the single traps, as it would be required in more complicated biological measurements.

In the past years, more technologies using microlens arrays, micromirror arrays, optical fiber arrays or even light patterned electrodes to create extended trap patterns have emerged. They lack partially a dynamic control of trap positions, but still provide useful tools for experiments in particle sorting and parallel detection approaches [78–80].

The methods mentioned until now, only use the amplitude and up to a certain degree the polarization of the incident light. But the properties of a light beam depend also on the phase and one of the advantages of a laser beam is that it can be seen as a very coherent light source. This means, that the wave vectors of the electromagnetic field have a constant relative phase, they are Òin-phaseÓ. Now, we want to introduce techniques that make use of this property to create multiple traps by interference effects.

One way to create interferometric optical traps is to bring two or more separate beams into interference [81–83]. With this method, extended two-dimensional and also three-dimensional trapping patterns have been achieved. Other methods induce interference by spatial modulation of the phase of an incident wave. The Talbot effect (Henry Fox Talbot, 1800-1877) is one possibility to create such an interference pattern. It is a near-field diffraction effect arising when a laterally periodic wave distribution is incident upon a diffraction grating. At regular distances, called the Talbot length, the image of the grating is repeated, and each of the points of the original grating can act as an optical trap. Three-dimensional trapping patterns and even atomic trapping have been achieved already using "Talbot optical traps" [84, 85]

Similar to phase contrast microscopy, it is also possible to convert a phase pattern into intensity gradients in the focal plane. This technique is called Generalized Phase Contrast (GPC) and uses computer generated dynamic phase patterns to generate tunable trap arrays [86, 87].

Finally, Holographic Optical Tweezers (HOT) use phase modulation of an incident laser beam to create multiple diffracted sub-beams by interference. Usually, interference patterns of a parallel light beam only occur in infinity, but a microscope objective in the beam path creates the Fourier Transform of the incident beam, thus creating the interference pattern, i.e. the optical traps in its focal plane [19, 21, 88, 89].

Due to their versatility and the possibility to create extended dynamic trapping patterns of reasonable strength, HOT have been used in this work. In the following chapter, we want to present the physical background of this technique more in detail.

2.6.2 Theory of Diffraction and Holography

If two different waves with amplitudes A_1 and A_2 and identical wavelength λ are superposed, then their amplitudes are combined to a new amplitude $A_3 = A_1 + A_2$. This effect is called interference. Depending on their relative phase shift this interference can be constructive or destructive to result in a new amplitude between 0 and $|A_1 + A_2|$, i.e. the amplitude is amplified or canceled out. Interference effects are observable in matter waves, sonic waves and also electromagnetic waves.

Holography is a technique that is based on interference effects between light beams, which are diffracted by an object. It was invented by the Hungarian physicist Dennis Gabor (1900-1979) who received the nobel prize in physics in 1979. Most people know holography as the method to reconstruct a three-dimensional image of an object by interference between scattered light and an undisturbed reference ray. Therefore, the light beam must be split up into two coherent subbeams. When now one of these beams passes the object, its phase will be shifted. Once the diffracted and the reference ray are combined again, an interference pattern will result. The interference pattern can be recorded for example on a photo sensitive film. This pattern contains the whole three-dimensional information of the object and it is possible to reproduce the three-dimensional image again by illumination of the interference pattern with the reference beam. Nowadays, holography is used for example in art and security applications, but also in three-dimensional imaging technologies as digital in-line holography [90].

In principle, it is also possible to obtain a three-dimensional image of an object by creation of the appropriate interference pattern artificially. Therefore, the required phase and amplitude distribution in the image plane must be computed. Holographic optical tweezers are based on this principle. The image to be reconstructed, is a distribution of light intensity in the focal plane of the objective. This distribution consists of points of high intensity, each of which represents one single optical trap. Since for trapping purposes only the amplitude of the light in the focal plane matters, the phase information can be neglected. So, the interference patterns that can create this amplitude distribution in the focal plane have a high degree of degeneration. Therefore, in practice phase-only holograms are used in order to minimize absorption and correspondingly intensity loss in the interference plane [19, 91]. Adaptive-iterative algorithms can be employed to calculate holograms with optimized efficiency [92].

To find the suitable interference pattern that creates the desired trapping array in the focal plane it must be considered that the objective or any lens in a light beam creates mathematically a Fourier Transform of the incident phase and amplitude information. This allows us to calculate interference patterns via an inverse Fourier Transform.

2.6.3 Fourier Optics

Light can be viewed as electromagnetic wave consisting of amplitude A and phase φ. The electromagnetic field of an incident wave at the locus $\vec{r}$ can be characterized as:

$$E^{in}(\vec{r}) = A^{in}(\vec{r})e^{i\varphi^{in}(\vec{r})}. \tag{2.46}$$

A planar array of optical tweezers can be described by the intensity distribution, $I^f(\vec{p})$ with the spatial coordinate $\vec{p}$, of laser light in the focal plane of a microscope's objective lens. Thus, the corresponding electric field in the focal plane has the form:

$$E^f(\vec{p}) = A^f(\vec{p})e^{i\varphi^f(\vec{p})}. \tag{2.47}$$

The intensity distribution $I^f(\vec{p})$ can be written as a function of the squared amplitude of the electric field:

$$I^f(p) \propto |E^f(\vec{p})|^2 = |A^f(\vec{p})|^2 \tag{2.48}$$

Both electric fields are related by the Fourier Transform $\mathcal{F}$ as depicted in figure 2.9:

$$\begin{aligned} E^f(\vec{p}) &= \mathcal{F}\{E^{in}(\vec{r})\} \\ &= \frac{k}{2\pi f}e^{-i\vartheta(\vec{p})} \int d^2r\, E^{in}(\vec{r})e^{ik\vec{r}\vec{p}/f} \end{aligned} \tag{2.49}$$

and

$$E^{in}(\vec{r}) = \mathcal{F}^{-1}\{E^f(\vec{p})\}$$
$$= \frac{k}{2\pi f} \int d^2p \, e^{-i\vartheta(\vec{p})} E^f(\vec{p}) e^{i k \vec{r}\vec{p}/f}. \qquad (2.50)$$

Here, k is the wave number of the incident light, f is the focal length of the employed

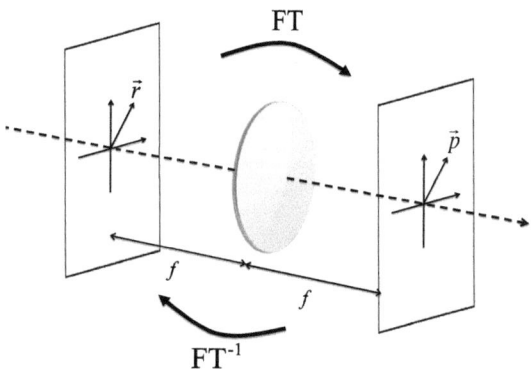

Figure 2.9: Relation between the electromagnetic fields in the focal plane and the back focal plane of a lens. The Fourier Transform of the incident light field is projected into the focal plane.

lens and $\vartheta(\vec{p})$ is an additional phase profile due to the lens' geometry. Since the phase profile is not relevant for the intensity distribution $I^f(\vec{p})$, the lens geometry can be ignored and the calculated holograms are universally applicable. A real example of a phase mask encoding a hexagonal intensity distribution in the focal plane is shown in figure 2.10.

In the following sections, we want to concentrate on the different ways to generate the desired intensity patterns and resulting trap arrays in the experiment.

2.6.4 Computer Generated Holograms

In experiments it is necessary to modify the incident wave of the trapping laser to provide the desired electric field E^{in}. As shown in the previous section, this is

2.6 Holographic Optical Tweezers

 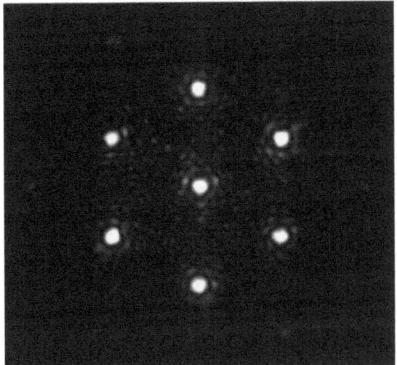

Figure 2.10: A phase mask encoding a hexagonal trap pattern seen on the right side. The grey levels in the hologram represent phase shifts between 0 and 2π. The phase mask is the Fourier Transform of the desired intensity distribution. The image on the right is a real image of the laser light distribution after diffraction at phase mask.

achieved by phase modulation of the wave front. There are different technical solutions that can act as diffractive optical elements, changing the phase of an incident wave front spatially resolved. This includes specially etched phase plates but also spatial light modulators, which are based on liquid crystals working either in transmission or reflection mode. Such a spatial light modulator (SLM) has been used in this work, which is the reason, why we want to concentrate on this technology now. A typical SLM consists of a liquid crystal display where each pixel can be independently addressed leading to phase shifts, ranging from 0 to 2π. A scheme of the operating mode of such a liquid crystal SLM and an exemplary SLM setup are presented in figures 2.11 and 2.12.

The pixelation of the phase modulator restricts the number of possible trap configurations, which limits the universality of the trapping patterns. A modulator with $M \times M$ pixels can therefore produce only M^2 discrete trapping positions after the discrete Fourier Transform. When aiming for dynamic adjustable holograms, this limits the minimum step length between two holograms, which can influence the measurements and lead to artifacts. The possible resolution δ_h of the hologram can be calculated from the pixel size of the modulator δ_m, the wavelength of the laser λ

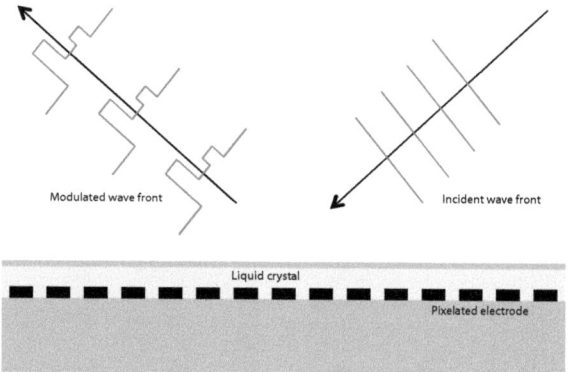

Figure 2.11: Schematic drawing of a liquid crystal phase modulator in reflection mode. The refractive index and therefore the optical path length inside the liquid crystal can be modulated by an applied voltage for each pixel. The phase of the incident wave front coming from the right is then modulated spatially resolved after reflection at the crystal.

and the specifications of the objective. If the objective has an aperture d, then we can assume that in the optimum configuration, this will completely be filled by the image of the SLM. We can write accordingly:

$$d = M\delta_m. \tag{2.51}$$

From the Fourier Transform (2.49) we get the relation between the pixel size in the focal plane δ_f corresponding to the respective pixel size δ_m on the phase modulator:

$$\delta_f = \frac{\lambda f}{M\delta_m} = \frac{\lambda f}{d}. \tag{2.52}$$

Here, f is the focal length of the objective. The relation between d and f can be expressed by the numerical aperture NA of the objective:

$$NA = \frac{d}{2f}. \tag{2.53}$$

So we can write

$$\delta_f = \frac{\lambda}{2NA}. \tag{2.54}$$

2.6 Holographic Optical Tweezers

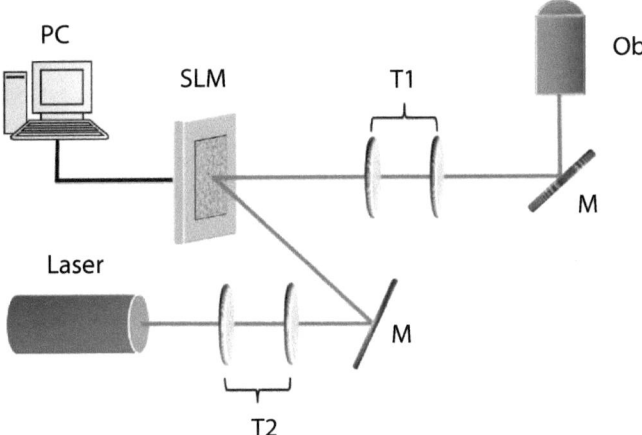

Figure 2.12: The diffractive optical element has to be placed into the back focal plane of the objective. Since usually this is not possible, it is placed into a conjugated plane. A telescope is employed to adjust the size of the diffraction pattern that way, that its image slightly overfills the input pupil of the objective. A second telescope enlarges the diameter of the incident laser in order to illuminate the whole SLM homogeneously.

It is now possible to determine the maximum resolution obtainable in the setup that was used in this work. An IR laser with a wavelength of 1064 nm and a water immersion objective with a numerical aperture of 1.2 were employed. This leads to a resolution of about 440 nm. An actin filament has a diameter of only 8 nm and a periodicity in its helix of 36 nm; cross-linker molecules have a typical size between 20 and 80 nm. It is obvious that the resolution of the Fourier Transform is an order of magnitude to low for dynamic measurements with these systems.

Nevertheless, one has always to remember, that this limitation is only a consequence of the discrete FFT used to calculate the holograms. By avoiding the discrete FFT, the lateral spatial resolution can be greatly increased. Schmitz et al. [93] developed an algorithm based on the superposition of electrical fields forming prisms and lenses which allows for the position steering of HOT with single nanometer resolution.

Such a prism can be seen as phase grating with an increasing phase shift which results in a displacement of the optical trap off-axis. In principle this would be the same effect, if a real prism was placed into the beam, letting the laser enter the objective under an angle. A phase profile can have the form:

$$\varphi(\vec{r}) = \frac{2\pi}{\lambda f} \left(\frac{f_1}{f_2}\right) \vec{p} \cdot \vec{r} \bmod(2\pi), \tag{2.55}$$

where (f_1/f_2) is the magnification by the telescope between the SLM and the objective; $\vec{r}$ and $\vec{p}$ are the coordinates in the SLM plane and the focal plane, respectively. The modulo function reduces the possible outcome to the values between 0 and 2π, which are addressable by the SLM. The steeper the phase profile is, the further the resulting trap is displaced from the focal spot of the objective. Analogously to equation 2.50, the electrical field of this hologram has the form:

$$\epsilon^{in}(\vec{r}) = \alpha^{in}(\vec{r})e^{i\varphi^{in}(\vec{r})}, \tag{2.56}$$

with $\alpha^{in}(\vec{r})$ being the respective amplitude. Such a phase profile and the resulting shifted trap in the focal plane are depicted in figure 2.13. The maximum step resolution for such a prism is defined by the number of pixels M of the SLM and the resolution of the phase shift of the liquid crystal g, which is the number of discrete steps between 0 and 2π. It is a fraction of the resolution for the discrete Fourier Transform because also non integer step functions are obtainable. Schmitz could show that for a system with M=512 pixel and g=130 phase steps, the resolution

2.6 Holographic Optical Tweezers

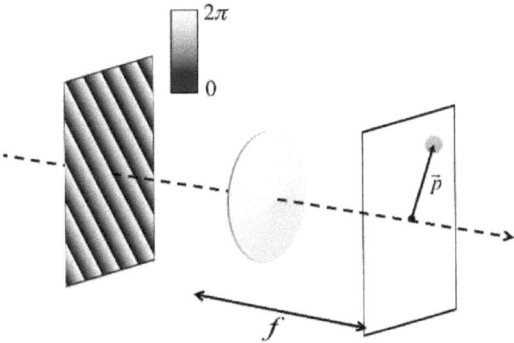

Figure 2.13: Prism hologram displacing an optical trap in the focal plane to a point $\vec{p}$.

easily reaches subnanometer values. Thus, the system is limited only by the localization of the traps via image analysis and external noise.

To calculate a certain phase profile for a desired trapping position we use the vector $\vec{r}$ in the aperture plane of the objective, which is conjugated to the SLM's plane, thus we can neglect the magnification factor (f_1/f_2) of the telescope. Every point in the SLM plane can be defined as: $\vec{r} = \delta_m(m\vec{e}_x + n\vec{e}_y)$. Replacing now the focal length in Eq.: 2.55 using the relation $f = M\delta_m/2\text{NA}$ from Eq.: 2.52, we can find the required phase profile for every trap position $\vec{p} = (p_x\vec{e}_x + p_y\vec{e}_y)$ in the focal plane:

$$\varphi(m,n) = \frac{4\pi\text{NA}}{\lambda M}(mp_x + np_y) \bmod(2\pi). \tag{2.57}$$

Furthermore, multiple traps can be created and individually controlled by superposition of multiple prism holograms, j. The electrical field of any pattern, containing N different traps writes thus as:

$$\begin{aligned} E^{in} &= \sum_{j=1}^{N} \epsilon_j^{in}(\vec{r}) \\ &\equiv A^{in}(\vec{r})e^{i\varphi^{in}(\vec{r})} \end{aligned} \tag{2.58}$$

But this would not be a pure phase hologram anymore, since the field E^{in} here is also dependent on the amplitude $A^{in}(\vec{r})$. However, it was shown that the amplitude information could be neglected and when using only the phase information $\varphi^{in}(\vec{r})$ satisfying results comparable to iterative algorithms could be achieved [93].

Using this superposition principle, single traps can be translocated, just by updating their fraction $\epsilon_j^{in}(\vec{r})$ of the original field $E^{in}(\vec{r})$. This reduces computation time for the calculation of holograms significantly and allows the dynamic control of trapping positions during the experiment.

Moreover, it is also possible to move single traps also in the z-direction by superposing phase holograms that have the form of Fresnel lenses (Augustin Jean Fresnel 1788 -1827) of the form

$$\varphi(\vec{p}) = \frac{2\pi p^2 z}{\lambda f^2} \mathrm{mod}(2\pi) \qquad (2.59)$$

where z is the desired displacement of the optical traps relative to the focal plane. The method of the prism superposition was employed in this work to calculate trapping patterns. In the presented experiments, the holograms were calculated in advance and saved as an image series to provide predefined trap motion in order to make single experiments comparable. However, due to reduced computing requirements the application of the superposition principle allows also for the *in situ* generation of holograms during the experiment.

Chapter

3

Actin and the Cytoskeleton

Just like higher multicellular organisms, cells have skeletal structures, which are responsible for their structural integrity and rigidity. But unlike in animals, on the cellular level this skeleton consists of filamentous structures made of proteins. This cytoskeleton is not a fixed structure but a highly adaptive and motile system, which interacts with its surrounding in a multitude of ways [1, 94]. Filaments are continuously polymerized and depolymerized, they are cross-linked by different proteins to form networks and bundles and they connect different cellular compartments. Additionally, the cytoskeleton has also sensory functions for the cell, it is used to probe the mechanical properties of the environment and is involved in cell signaling. Moreover, the cytoskeleton is a railway system for the cell; motor proteins transport cargo along these tracks overcoming diffusion limitations when distributing reactants to different locations inside the cell. Furthermore, the cytoskeleton plays a crucial role during cell motion, division and differentiation [2, 95–97]. It is formed mainly by three types of proteins, microtubules, intermediate filaments and actin.

3.1 Cytoskeletal Components

3.1.1 Microtubules

Microtubules (MT) are hollow tubes, made of 13 protofilaments. They have a diameter of 25 nm and consist of the protein tubulin, a heterodimer with the subunits α-tubulin and β-tubulin. The monomers have a molar mass of about 50 kDa. The filament is polar having a minus and a plus end with different polymerization kinetics and binding affinities to associating proteins. In most cells the microtubules are oriented with their plus end pointing to the rim of the cell while the minus ends are located near the nucleus. During mitosis, MT nucleate from microtubule organizing centers, the centrosomes, to form the mitotic spindle, which separates the chromosome pairs [1].

Two motor families use microtubules as track: kinesins which transport cargo always towards the plus ends of the MT and dynein that walks towards the other direction. Dynein is also responsible for the power stroke in flagellae [98]. MT are also important for internal organization and mechanical rigidity of cells. Due to their high rigidity, they have a persistence length of more than 1 mm [99, 100], and are therefore believed to sustain compressive loads in cells [101]. MT are interconnected with other cytoskeletal compartments and it is proposed that the induction of stress and compression between actin and microtubules supports the structural integrity of the cells according to the tensegrity model proposed by Ingber et al. [2, 102].

3.1.2 Intermediate Filaments

Intermediate filaments (IF) are much less conserved than actin and microtubules. IF are a family of over 50 different proteins and their individual properties are less investigated than the other cytoskeletal components. The most prominent IF proteins are kreatin, vimentin, desmin, nestin and the nuclear lamins. They form filaments with a diameter of about 10 nm which is between the size of actin filaments and microtubules, why they are given their name [1]. Each filament is constructed from a series of subfilaments that are connected to each other at their ends. In the cytoplasm, they are distributed from the nuclear surface to the plasma membrane and they also form networks in the nuclear envelope. IF are connected to the rest of the cytoskeleton via associating proteins including desmoplakin and plectin. They are also involved in cell-surface and cell-cell junctions, the hemidesmosomes and desmosomes which are crucial for tissue integrity. There are no known motor proteins to walk along intermediate filaments. IF are very resistant to stress and their important role for the mechanical strength of cells was demonstrated in keratin knock-out experiments with epithelial cells [103, 104]. It could be shown that IF are crucial for the cells to maintain their shape and the mechanical integrity of the cytoplasm [105]. They are also prominent in the axons of neurons, where they form the neurofilaments.

3.1.3 Actin

Actin filaments are also known as microfilaments since actin is the smallest filament in the cytoskeleton. It is the most abundant of all eukaryotic proteins and also, one of the most conserved. In non-muscle cells, more than 25 % and in muscle cells more than 60 % of the protein mass is actin [98]. Actin is one of the most investigated proteins. There are more than 60 different classes of actin binding proteins and still new ones are found. It is important for all types of cellular motility, including cell division, exo- and endocytosis, lamellipodium protrusion, amoeboid motion and muscle contraction [106–112]. It regulates the viscoelastic properties of the cytoplasm, it forms the mechanic support for the cell membrane and it is involved in the sensing and regulation of stress and tension on the cell [2, 102, 113–116]. Actin forms an extended track system for cargo transport by motor proteins similar to microtubules [117].

Actin structures do not only feel mechanic stimuli but react to them and remodel their organization and shape in response [118]. Therefore, mechanical and chemical properties of the protein are heavily interconnected and represent a highly interesting topic for research. In this work, we tried to investigate how actin produces forces during cross-linking and how such cross-links withstand applied forces. Thus, we we will look into the details of different roles and appearances of actin in cells.

3.1.4 Biochemistry and Biology of Actin

Actin is a highly conserved polypeptide of 42 kDa molecular mass. It consists of 4 subdomains that enclose a nucleotide binding cleft. In cells, the globular monomer G-actin is polymerized to form the filamentous F-actin. The filaments have the form of a double stranded helix with a diameter of 5-9 nm and a periodicity of 36 nm [119–122]. Actin filaments have lengths in the order of tens of microns and, *in vitro*, also lengths of up to 100 μm can be observed (see also figure 7.1). The persistence length over which correlations in the direction of the filaments tangent are lost is for actin in the order of 17 μm[123]. Being a measure for the flexural rigidity it is in the same order of magnitude as the filaments contour length, classifying actin as a semiflexible polymer [36, 99, 124]. Therefore, in cells actin is often found cross-linked or bundled to increase its mechanic strength. Actin undergoes only a small number of post translational modifications. Typical modifications are acetylation of the N-terminus of the protein and methylation at His 73. Interestingly, this methylation is missing in *Plasmodium* actin, which might explain the very short filament lengths observed [125]. Polymerization of actin filaments is a multi-step process. A schematic of the different steps involved is shown in figure 3.1. Polymerization starts with the formation of an oligomeric nucleus. To form the nucleus, the electronic repulsion of the monomers that each bear a charge of $11\,e^-$ has to be overcome. This nucleation step is very slow since the binding energy that can be gained by this process is relatively small. In an actin filament, each monomer can have interactions to 4 neighboring partners, while in the polymerization nucleus it interacts only with one or two partners. Therefore, this step is the rate determining step. Filament growth can happen on both ends of the fiber. The actin monomer itself is asymmetric and accordingly also the filament has a directionality, a so-called barbed and a pointed end. Adsorption on both ends will follow different kinetics, which means that different equilibrium constants exist. Since the concentration of monomers in solution is the same for both ends, it can happen that one end still grows, while the other is shrinking due to the reduced monomer concentration. When shrinking at one end and growth at the other have the same rate, a dynamic equilibrium is reached. Filament length and monomer concentration in solution do not change anymore, but monomers in the filament are "transported" from the growing to the shrinking end. This state is termed treadmilling and the average velocity of the monomer migration in the filament during this process is $2\,\mu m/h$. So, for actin polymerization four concentration regimes can be defined. A low concentration regime below a critical concentration where the disassembly of nucleation cores is faster than the the formation and no filaments are found in solution. A steady state concentration where the filament is treadmilling and two intermediate states, either characterized by growing or shrinking filaments. Only the first two states are stable since the latter two will always lead to one of the other two states. The rates of nucleation and elongation are strongly dependent on the medium conditions. ATP is required for polymerization and high ionic strengths and the presence of divalent cations promote this process [126, 127]. Cells can actively control the filament assembly inside the cytosol by different actin binding proteins (ABP). These ABP prevent monomers from assembling by complexation, cap ends of growing filaments, stabilize nucleation cores to promote filament formation and they can also sever existing filaments to provide new nucleation centers [128–131]. Apart from influencing the

3.1 Cytoskeletal Components

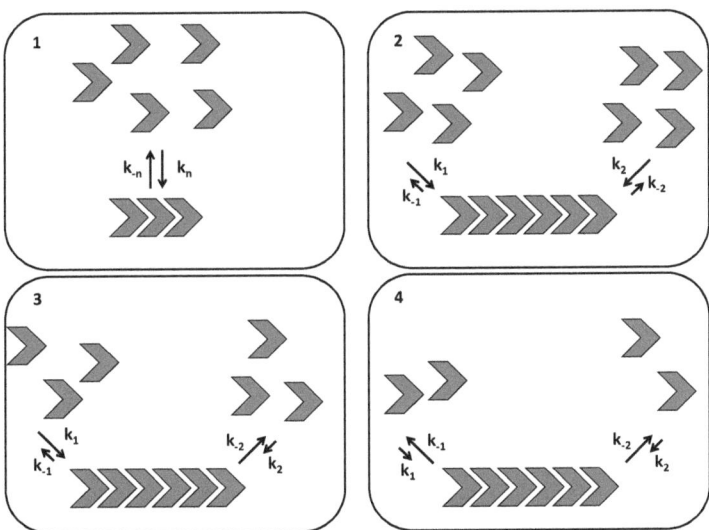

Figure 3.1: The figure shows a schematic summary of the mechanism of actin polymerization. 1: Actin monomers have to form nucleation cores, defined by a rate constant k_n. Nucleation happens on both ends following different rate constants k_1 and k_2. At high monomer concentrations the filament grows on both ends (2). At intermediate concentrations the filament still grows on the barbed end but shrinks already at the pointed end, this state is termed the treadmilling state (3). At low monomer concentrations the filament will shrink on both ends and become shorter (4). Cells can actively regulate the pool of available monomers by complexation to tune their actin polymerization behavior.

polymerization kinetics, cells have a wide range of other possibilities to regulate their actin cytoskeleton. Actin binding proteins can interconnect different actin filaments to form bundles or networks. They can also form connections between the actin filaments and other intracellular compartments like microtubules, intermediate filaments or membrane compartments [132–134]. Finally, there is a special class of actin binding proteins that can exert forces on filaments. These molecular motors use actin as a trail or can actively move two filaments against each other by converting chemical energy during ATP hydrolysis[135].

3.1.5 Actin Binding Proteins

In this work, we employed two different actin binding proteins: α-actinin an, actin bundling protein and myosin II, the most abundant molecular motor which is responsible for muscle contractility. Inactivated myosin was used to bind actin to pillar tops in the assembly of two-dimensional networks on PDMS pillar substrates and α-actinin was investigated as bundling reagent. Therefore, we want introduce these two molecules briefly.

Myosin II

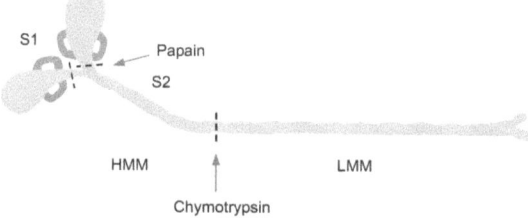

Figure 3.2: The figure shows a schematic model of a myosin II molecule. Heavy peptide chains are depicted in green and orange. The light chains are colored blue. Chymotrypsin cleaves the coiled-coil into LMM and HMM residues. By enzymatic degradation with papain, the head subfragment S1 can be obtained. Graphic courtesy of Christian Schmitz [136].

Myosin II belongs to the myosin motor protein family. All myosin molecules are known to bind to actin and to generate force by ATP hydrolysis to perform a walking motion on the filaments. Myosin II, also sometimes called conventional myosin, was the first molecular motor which was discovered. It is a hetero hexamer that is built by 2 heavy peptide chains of about 200 kDa and four light chains of about 20 kDa. It consists of two globular heads and a long coiled-coil that are connected by a neck region (see figure 3.2). Actin binding sites as well as an ATP binding center are located in the head regions. Myosin II is found in high concentrations in the sarcomeres of muscle tissue [111]. The tail regions of myosin form thick bundles in muscle, while the head portions of the protein reach out from the core to interact with adjacent actin filaments. By the orchestrated interaction of millions of myosin

3.1 Cytoskeletal Components

heads with actin fibers the contractility of muscle fibers is produced. The C-terminal part of the tail domain, the light meromyosin (LMM) can be cleaved by the enzyme chymotrypsin from the rest of the molecule, the heavy meromyosin (HMM) [137]. The remaining part of the myosin is soluble at physiological conditions. Using the protease papain, also the head region can be separated from the residual chain [138]. Treatment of HMM with N-ethylmaleimide (NEM) produces an inactivated myosin that looses its motor activity and remains bound to actin even in the presence of ATP [139].

α-Actinin

α-actinin was first isolated from muscle tissue as a component of the Z-disc, but was later also found as a component of focal adhesions and stress fibers in adherent cells [140, 141]. It is a rod-shaped heterodimer with a molecular subunit mass of 103 kDa. The monomer consists of a globular N-terminal actin binding domain, four spectrin repeats and a pair of calcium binding EF-hand protein motifs. In the dimer

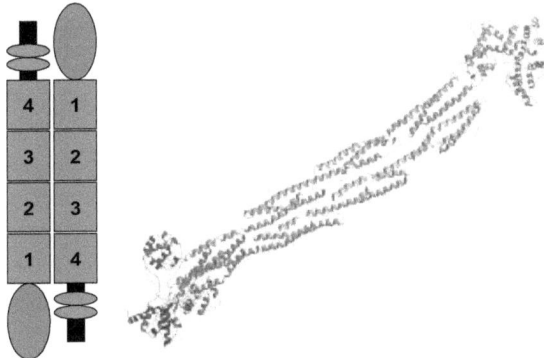

Figure 3.3: Peptide motifs of α-actinin. The actin binding site is colored in green. Spectrin subunits are green and the EF motif in blue.

the molecules are arranged antiparallel. The binding is mediated by the spectrin subunits. The α-actinin molecule is between 30 and 44 nm long and 7 nm wide. It can bundle two parallel actin filaments by its two binding sites, but also networks of actin cross-linked by α-actinin can be observed [142–144]. Cells are highly dynamic and, hence, persistent cross-links clearly would be disadvantageous. Therefore, one can expect, that the detachment of α-actinin happens at relatively low forces to allow the cells to continuously remodel their shape and structure. Recent experiments by Miyata *et al.* and Ferrer *et al.* showed a rather broad distribution of possible cross-linking forces for α-actinin also including high forces ranging from 1.4 pN up to 80 pN [37, 38]. However, these experiments were all performed with surface bound actin filaments. In freely fluctuating filaments additional effects due to thermal motion should facilitate the rupture of the cross-link [35]. Surface effects like unspecific binding of the probe could lead to additional artifacts. Therefore, methods that could measure such events in the more natural environment of freely suspended filaments were developed during this work.

3.1.6 Actin Cortex and Actin Bundles

In cells, actin can be found in network structures or in thick bundles, the stress fibers. Also intermediate forms, networks of bundles, are possible [113, 144, 145]. The actin cortex is a network of entangled, partially cross-linked filaments located directly underneath the cell membrane. It is connected to the cell membrane by actin binding proteins that can transduce signaling to the cortex and remodel its structure [108, 146]. Being a highly adaptive and regulated network, it plays a crucial role in the interaction of the cell with its environment. It is responsible for the structure and the shape of the cell and has to withstand external deformations to provide elasticity and stability to the cell. Its properties are strongly influenced by its confined structure. Even though it is extended over the whole surface of a cell, its thickness is very small. It only consists of a few monolayers of actin filaments reaching several hundred nanometers into the cytosol [31]. Compared to the contour length l_c and the persistence length l_p of actin filaments which is in the order of tens of micrometers, it can be considered as a quasi two-dimensional structure. Actin networks are solutions of entangled semiflexible polymers. Thus, they show high frequency dependent viscoelastic characteristics. Different models have been developed to describe the unique properties of this biopolymer structures that are still not matched by synthetic materials. The reptation model for example explains the confinement of single filaments by virtual tubes representing the surrounding filament network to describe the degrees of freedom for entangled filaments [147, 148]. However, it is not yet understood how such models scale when restricted to two-

3.1 Cytoskeletal Components

dimensional systems. Thus, experimental techniques that allow the construction of two-dimensional actin networks to investigate their mechanical properties and their behavior during cross-linking were developed to tackle these questions.

Actin bundles are the major players in muscle contractility. Furthermore, they are responsible for the formation of cellular protrusions, the microvilli and they also play an important role in transport processes inside cells [117, 149]. In actin bundles, multiple filaments are aligned parallel and tightly packed. Actin bundling proteins accomplish this process overcoming the electrostatic repulsion of the negatively charged actin filaments [150]. The bundles are not permanent structures inside cells. Instead, they are highly dynamic, they form and dissolve continuously and change their form, thickness and shape [151]. Therefore, the bundling properties of the different cross-linking molecules have to be adjustable and tunable in order to allow cells to control their structure.

To form an actin bundle, two filaments have to come to close proximity and a cross-

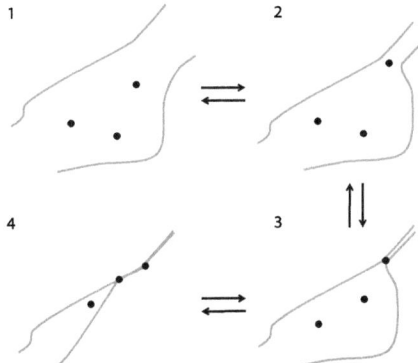

Figure 3.4: Process of zipping of two actin filaments (green) by cross-linking molecules (black dots). Every step is in an equilibrium which is defined by thermal motion, binding affinity, concentrations and external load.

linking molecule has to be present at the binding site to enable the connection. But this attachment is only transient. The filaments undergo continuously Brownian motion, which could open the cross-link again. But due to the initial cross-link, the average distance of the filaments is reduced. The chance of a new cross-link in close proximity to the first binding point is increased accordingly. If enough cross-linking molecules are present in the solution, the two filaments will close in a zipper like way to form a bundle [152]. The different basic steps of this process are shown in figure 6.1. In this work, we wanted to find methods to quantify the relevant forces of this zipping process. Therefore, biomimetic models that involve only the minimum required partners, cross-linking molecules and two filaments in the right geometry, were developed. We used optical tweezers and PDMS micropillar substrates in microfluidic environments as scaffolds for the zipping structures to perform experiments that should give insights into the zipping of two filaments as well as into the forced unzipping process by an applied external force.

Chapter

4

Malaria and Biophysics of the Malaria Parasite

Malaria is still one of the leading causes of death in the tropics [153, 154] and is likely to influence the life in northern countries in future more and more due to changes in the global climate [155–157]. It is a vector-borne infectious disease caused by a protozoan parasite, the *Plasmodium* parasite. Usually, the parasite is transmitted to humans by the bite of an infected mosquito of the genus *Anopheles*. The *Plasmodium* parasite has a complex life cycle that involves different stages in vertebrate and insect hosts. During this life cycle, the parasite has to adapt to different intra- and extracellular environments, interact with different tissue and use a unique form of locomotion, called gliding motility to target and invade host cells [158, 159]. Gliding motility is an actin based movement that is unique to apicomplexan parasites. Apicomplexa are an ancient protozoan phylum, that includes also *toxoplasma*, a parasite that can have fatal effects on the unborn live and *babesia*, which is one of the most common blood parasites of mammals, but luckily not very common in humans [160–162]. A better insight into the molecular machinery of gliding might provide targets for drug application, which block the migration and adhesion of the parasites and therefore prevent infection and disease. In this work, we wanted to investigate the of mechanism parasite adhesion eventually leading to productive gliding. Moreover, we aim to elucidate, which proteins are involved in this multi-step process. The optical tweezers as an instrument to manipulate and probe objects at the microscopic scale were chosen as an ideal tool to investigate the adhesion of the sporozoite stage of the *Plasmodium* parasite. This part of the work was performed in close collaboration with Stephan Hegge from the group of Friedrich Frischknecht at the department of parasitology, University of Heidelberg.

4.1 History and Background of Malaria

Malaria is known to infect people for more than 6000 years but estimations based on genetic analysis revealed that the malaria parasite might plague humans already for up to 100.000 years [163]. The malaria parasite was first observed 1880 inside the red blood cells of infected people by the French army doctor Charles Louis Alphonse Laveran (1845-1922) who was awarded the Nobel Prize for Physiology or Medicine in 1907. The Anglo-Indian physician Sir Ronald Ross (1857-1932) who got the Nobel Prize for Physiology or Medicine in 1902 was the first to prove in 1898 that malaria is transmitted by mosquitoes. He was also the first who discovered the role of vertebrate and insect host in the life cycle of the malaria parasite.

In the beginning of the 20th century, the Austrian psychiatrist Julius Wagner-Jauregg, (1857-1940) infected patients that were suffering from mental diseases caused by neurosyphillis deliberately with malaria. The resulting fever led to an dramatic improvement in the original symptoms of the patients that survived this cure. This treatment became an accepted therapy for the syphilis caused general paresis of the insane. For this discovery he was awarded the Nobel Prize for Physiology or Medicine in 1927 as the first of only two psychiatrists ever [164]. Eventually, the discovery of antibiotics against syphilis made this harsh method inadequate and useless

Currently malaria is widespread in most tropical and subtropical regions of America, Africa and Asia and even in some parts of Europe. Friedrich Schiller got infected with malaria, when he was in Mannheim to perform the premiere of his piece Don Karlos in 1783 [165]. According to the World Health Report of the WHO, every year, more than 500 million cases of malaria are reported and 3 million people are killed, most of them children in subsaharan Africa [154, 166].

One of the first treatments for malaria was quinine, which was extracted from the bark of the chinona tree in South America. It is named after the Countess Anna de Chinchon; the wife of the Peruvian vizeroi who was treated in 1633 by the monk Antonio de Calancha with this medicine [167]. This treatment was already known for centuries amongst the inhabitants of Peru and was brought to Europe in the 17th

century. Quinine was widely used by the British Army in the form of tonic water as malaria prophylaxis. However, mostly mixed with other compounds to balance its extremely bitter taste.
The French chemists Pierre Joseph Pelletier (1788-1842) and Joseph Bienaimé Caventou (1795-1877) isolated and identified quinine for the first time in 1820 [168]. Nowadays, there are several drugs for the treatment and prophylaxis of malaria available, the most prominent are chloroquinine, quinine and artemisine, but a growing degree of resistances in *Plasmodium* parasites makes medication more and more ineffective and leads again to a rise in malaria mortality [169]. This creates a persistent need for research on novel therapies and strategies to fight this fatal disease.

4.2 Biology and Life Cycle of the Malaria Parasite

There are more than 200 species in the genus *Plasmodium* and only 4 of them cause malaria in humans. *Plasmodium falciparum*, *Plasmodium vivax*, *Plasmodium ovale* and *Plasmodium malaria*. In this work, parasites of the rodent specific strain *Plasmodium berghei* were chosen as a model system. They posses an almost identical motility system compared to the human pathogenic strains but pose no risk to man during lab experiments.
The life cycle of the *Plasmodium* parasite involves different stages in vertebrate (e.g. human) and insect (e.g. *Anopheles* mosquitoes) hosts. Between these stages, the parasites change their morphological appearance several times. Each stage is characterized by a distinct shape, biochemistry and gene expression patterns. In figure 4.1 a comprehensive scheme of the different stages is shown. A malaria infection starts with the injection of the *Plasmodium* parasites during a blood meal of the female *Anopheles* into the skin of a host. The parasites migrate though the skin tissue to eventually find and invade either a lymph vessel or a blood capillary to enter blood circulation [39, 170]. Sporozoites that made it into the blood circulation keep floating until they reach the liver [171, 172]. After migrating through several layers of hepatocytes they finally invade one liver cell by forming an paristophorous vacuole, transform and replicate to several thousand merozoites by means of asexual schizogeny. Tens of thousands of merozoites per invaded cell are finally released into the blood stream in vesicles named merosomes [173–175]. The time from sporozoite invasion to merozoite egress, the tissue phase, varies between 8 and 30 days and is

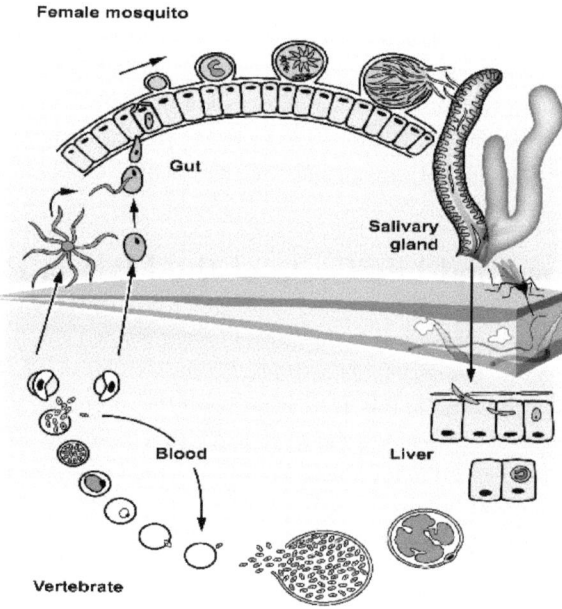

Figure 4.1: The life cycle of the *Plasmodium* parasite. The upper part shows the mosquito stages. Zygotes are in the gut of the insect and develop in the midgut wall into oocysts that subsequently release sporozoites, which travel to the salivary glands of the mosquito. From the salivary glands the parasite is injected into the blood vessels of the host. There, it migrates to the liver where it transforms into schizonts and finally invades blood cells in the merozoite form. During another step of schizogeny in this blood stage more merozoites are formed and thus build an inner loop within the whole life cycle. In some cases, schizogeny leads to production of sexual forms. Once these so called gametocytes are ingested by another mosquito they form zygotes. The life cycle is completed and starts all over again.

termed the prepatent period.

Each of the merozoites can invade an erythrocyte in a multi-step process [176]. Inside the red blood cell (RBC), a new asexual division cycle (schizogeny) starts, which results in proteolysis of the cellular hemoglobin. The cycle results in the release of around 20 merozoites per infected erythrocyte, which could invade further RBC. Fragments of dead RBCÕs are not flexible anymore and clog small capillaries. Furthermore, these pieces are recognized by the immune system leading to a global inflammatory reaction. This reaction is responsible for the characteristic symptoms of malaria, including the heavy fever and lasts for about 24 to 72 hours, depending on the malaria strain. A small part of the merozoites inside the RBC finally differentiate into the sexual form of the parasite, the micro- and macrogametocytes.

Gametocytes have no further interaction with the human host but are crucial for the transmission back to the insect host, the *Anopheles* mosquito. Once ingested by a mosquito, the macrogametocytes form macrogametes and the microgametocytes produce microgametes by exflagellation inside the midgut of the host. The gametes fuse and form a zygote after fertilization. This zygote transforms into an ookinete, which penetrates the wall of a cell in the midgut and develops into an oocyst. Inside the oocyst sporozoites are produced (sporolation) and released during rupture of the oocyst. The sporozoites migrate through the hemolymph and bind specifically to the salivary glands for onward transmission to a new host, where the life cycle begins again [177, 178].

4.3 Motility and Adhesion of *Plasmodium* Sporozoites

Motility is a crucial factor during the life cycle of the *Plasmodium* parasite and thus an important factor in the way of a malaria infection. In the presented experiments, we want to concentrate on the motility of the sporozoite stage, since this is the form with the most pronounced active motility in the *Plasmodium* life cycle. Sporozoites have to migrate to the salivary glands inside the mosquito. They move slowly within the salivary ducts and speed up upon transmission into the skin, where they can migrate through the dermis and enter blood or lymph vessels to transport them to the liver [39, 179]. Once arriving at the liver they are known to actively traverse several cells before they infect one. So the motility of the *Plasmodium* sporozoite has a multitude of facets, which could disclose novel targets for potential malaria drugs. The fact that 50% of all sporozoite proteins are unique to this stage in the life cycle of the parasite explains its importance in drug target research [180].

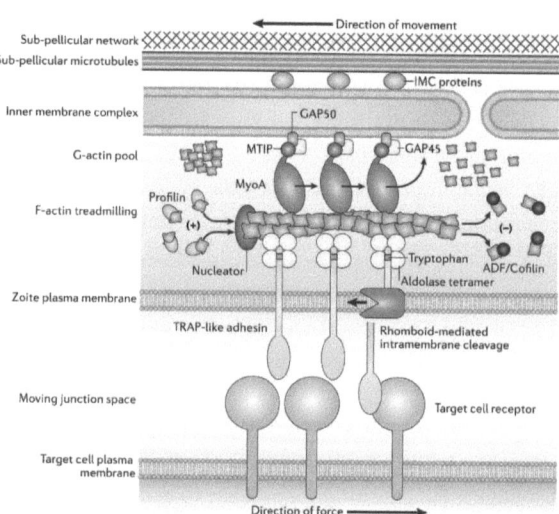

Figure 4.2: The figure shows a model for sporozoite gliding motility. Adhesins from the thrombospondin-related anonymous protein (TRAP) family are responsible for adhesion to substrate or host cells. A treadmilling actin filament regulated by different actin associated proteins is thought to be responsible for transduction of force. This force is produced by myosin, anchored to the inner membrane complex. Actin and bound adhesins are transported backwards, which leads to a forward motion of the parasite. Figure adapted from Baum *et al.* [159].

In general, it is possible to distinguish two major types of substrate associated motility. Crawling or amoeboid movement, in which obvious changes in cell morphology occur and gliding movement in which no changes in cell shape are observable. The first is typical for amoebae and many other cell types, e.g. fibroblasts or neurons [110, 181], while the second is found in trophozoites and sporozoites [182]. In contrast to amoeboid movement, sporozoite migrate in the absence of extended actin structures like filopodia. Instead very short fragments of actin filaments seem to be

4.3 Motility and Adhesion of *Plasmodium* Sporozoites

involved in the locomotion of sporozoites. Motility could be suppressed by cytochalasin D, an actin filament disrupting compound [183, 184]. Nevertheless, until now it was not possible to resolve actin filaments inside sporozoites by conventional fluorescence microscopy which is likely to be caused by their very short contour length of only 100 nm [125].

A model has been proposed, that involves the secretion of proteins to mediate cell adhesion during gliding as well as intracellular membrane compartments, the inner membrane complex (IMC) and an actin based machinery to provide motility (see figure 4.2) [159, 185]. The same motility system seems to be involved in gliding on substrates or in tissue as well as in the traversal of host cells by sporozoites in the liver [186, 187]. Members of the thrombospondin-related anonymous protein (TRAP) family of transmembrane proteins are known to form a connection between the extracellular adhesion and the intracellular actomyosin complex. The TRAP family proteins are unified by their transmembrane architecture. It is responsible for the coupling of host cell recognition by putative adhesive domains within the extracellular region to motility via a short charged cytoplasmic domain that interacts with the actin system [188, 189]. In *Plasmodium* at least six members have been found: TRAP, CTRP, PTRAMP, MTRAP, TLP and S6, out of which TLP (Trap-Like-Protein) and S6 (Sporozoite Induced Protein) are not fully characterized yet [190]. A comprehensive scheme of the domain architecture of the different proteins of this group is shown in figure 4.3.

Plasmodium Sporozoites have an elongated, curved structure. They are about 10 µm long and have a diameter of 2 µm. During *in vitro* studies, sporozoites show a circling motility behavior [171, 194], a locomotion in a circular path of 10 µm in diameter. Interestingly, nearly all sporozoites circle counterclockwise on the substrate. During the circling the sporozoites form multiple attachment sites to the substrate (Sylvia Münter, *unpublished results*). In addition to circling, a second form of motility has been reported, the so-called waving [171]. Waving sporozoites are attached on one end and actively move the second end, probably to search for a second adhesion site. An image sequence of this motility patterns can be seen in figure 4.4. Finally, there are some other forms of motility, which are partially induced by genetic defects or developmental immaturity. Patch gliding sporozoites for example are attached to a single site on the surface, but the whole sporozoite moves repeatedly over this adhesion site back and forth again, not being able to

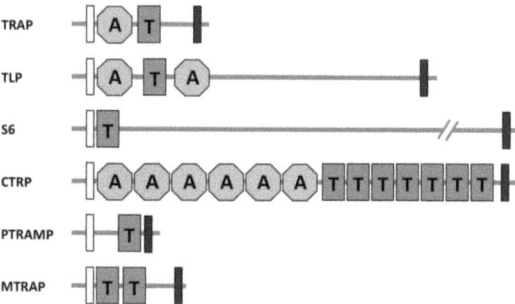

Figure 4.3: Schematic drawing of the domain architecture of TRAP family proteins found in *Plasmodium*. The first three, TRAP, TLP and S6, are mainly expressed in the sporozoite stage, CTRP is prominent in the ookinete form and PTRAMP and MTRAP in the merozoite form of *Plasmodium* [185, 191–193]. Relative protein lengths are drawn to scale. The white boxes are signal peptide sequences and the black boxes transmembrane domains. Green octagons, labeled with **A** are integrin like von Willebrand factor A domains and red boxes labeled with **T** represent thrombospontin type 1 domains [186, 189].

form stable secondary adhesion sites.

During *in vitro* experiments, a sporozoite performs multiple steps before motility

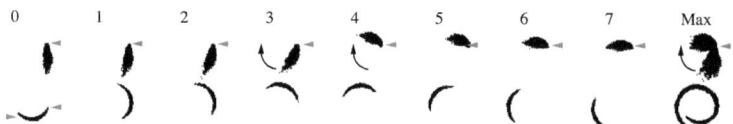

Figure 4.4: The figure shows an image sequence of the two prominent forms of *Plasmodium in vitro*. In the top row a waving sporozoite, attached at the point marked by green triangle. Maximum projection of all images on the right. Below a counterclockwise circling sporozoite. Images are micrographs, inverted and threshholded for better contrast, time lapse is 1 second between images. Courtesy of Friedrich Frischknecht.

can be observed. Usually, sporozoites are floating inside the solution in the beginning of the experiment. After approaching to the surface they establish a first contact and form an adhesion site. Initial adhesion is established randomly at a single site on the sporozoite. A secondary adhesion is commonly formed after translocation of the parasite over the adhesion site to adhere at the rear end of the parasite. This step is often anteceded by waving phases. Secondary adhesion can be induced by external forces such as laminar flow that bring the parasite close to the surface. It is followed by a slight stretching and flipping over of the sporozoite such that the complete parasite is in close proximity to the substrate (tertiary adhesion) and could initiate gliding (Stephan Hegge, *unpublished results*). It is still not known, which of this steps requires actin and which proteins are involved in the formation of the different adhesion sites. See figure 4.5 for a scheme of the different adhesion steps before gliding of the parasite.

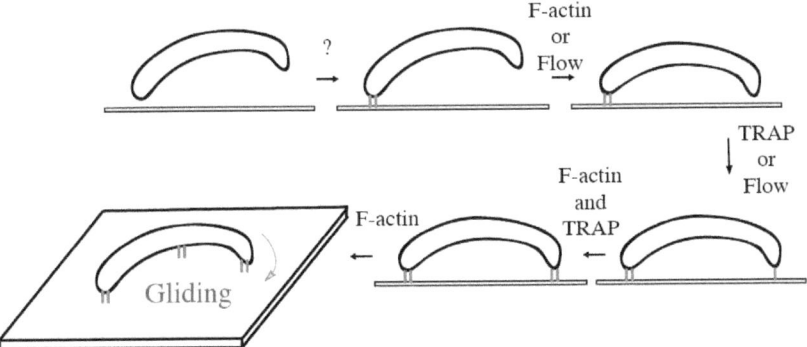

Figure 4.5: Schematic drawing of the multiple steps of sporozoite adhesion before circling. Secondary adhesion could also be induced by external forces such as viscous forces due to flow. The involved proteins are still not fully identified, but actin, TRAP and members of the TRAP family seem to be required.

In this work, we probe the adhesion of the sporozoites to surfaces using optical tweezers. Adhesion properties under the influence of actin disrupting drugs as well as adhesion of knock-out parasites, lacking relevant adhesion proteins are investigated to study the mechanism of motility and interactions with surfaces. A multi-step adhesion process is proposed, that requires the concerted interaction of different involved proteins that could partially substitute each other.

Part II

Materials and Methods

Chapter

5

Microscopy Setup

The microscopic setup employed in this work is a combination of three different, independent video microscopes with an additional holographic optical tweezers device coupled into the beam path. In the high magnification part of setup, trapped particles were imaged at high frame rates with nanometer resolution to allow the determination of pN forces. In low magnification, the microfluidic flow cells were imaged to control filling of the channels. Furthermore, by choosing suited beam splitters it was possible to image stained protein structures between the particles by fluorescence microscopy.

In the experiment, simultaneous imaging of these three sources had do be achieved without the use of mechanic shutters or mirrors, which could interfere with the force measurements. The microscope was based on an Alpha-Snom platform from Witec (Ulm) equipped with two oppositely situated confocal objectives. The tweezers setup and the beam paths for fluorescence microscopy and high-speed imaging were additionally implemented. The whole setup was placed on a damped optical table to be isolated from external vibrations. Furthermore, the part of the inverted microscope was embedded in a closed cage to minimize fluctuations in the surrounding air, which could disturb the trapping laser beam. The cage also served as protection against the invisible IR radiation of the trapping laser. A scheme of the complete setup used in this work is shown in figure 5.1

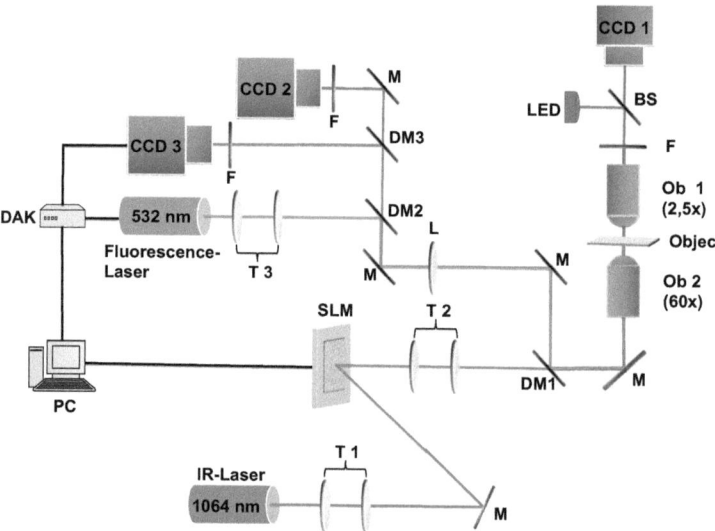

Figure 5.1: The complete microscopic setup used in the experiments. In the upper right corner the low magnification upright microscope (**Ob1**) with the 633 nm LED illumination and a CCD camera (**CCD3**) is shown.
On the left side is the inverted microscope including the optical tweezers and the fluorescence setup. The IR laser is used for the trapping. The first telescope (**T1**) expands the beam to overfill the spatial light modulator (**SLM**), the second telescope (**T2**) adjusts the beam size to cover the back aperture of the objective (**Ob2**) and places the SLM into a plane conjugated to the back focal plane of the objective. The dichroic mirror **DM1** transmits IR and reflects below 950 nm. The green laser (532 nm) is used for fluorescence excitation, it is focused into to the back focal plane of the objective to achieve a homogeneous illumination. The fluorescence excitation light (575 nm) is reflected by the dichroic mirrors **DM1** and **DM3** and transmitted by **DM2** to be recorded on **CCD1**. The high-speed camera (**CCD2**) uses the transmitted light of the LED, which is reflected by **DM1** and transmitted by the other two dichroic mirrors. All cameras are protected by suited filters from light of any unwanted wavelenths. The cameras and the fluorescence laser are synchronized via a data acquisition card (**DAC**) and computer controlled.

5.1 Holographic Optical Tweezers Setup

For the creation of the optical traps, a 5 W solid state NdYVO$_4$ laser (J20-BL-106C; Spectra Physics, Mountain View, CA) with a wavelength of 1064 nm was used. The wavelength of the laser was chosen to fit the transparency window of water and most biological samples. This prevented unwanted heating of the sample and limited photo-damage of trapped specimen to a minimum. The laser was pumped by a diode laser and via adjustment of the diode current, one can tune the output power of the trapping laser. It is working in the TEM$_{00}$ mode and the output light is planar polarized.

The phase of the laser was modulated by a 512 x 512 pixel nematic liquid crystal light modulator (512x512 SLM System; Boulder Nonlinear Systems, Lafayette, CO). The molecules of the crystal were anisotropic in their refractive index. By application of an electric voltage it was possible to change their orientation, which allowed to change the phase of the reflected light for each pixel. The SLM provided 130 distinct phase shifts between 0 and 2π.

The laser was focused into the object plane by a water immersion objective (Universal Plan Apochromat 60x, NA 1.2 W3 IR; Olympus, Center Valley, PA). This objective had a relatively high transmission in the IR ($> 60\%$) and a high degree of aberration correction. Compared to oil immersion objectives, having a higher numerical aperture, allowed the use of an water immersion objective trapping also at larger distances from the surface. Due to refractive index mismatch between the immersion medium and the sample medium optical traps from oil immersion objectives are stable only over several micrometers depth inside the sample. With the water immersion objective, stable trapping could also be achieved at depths above 50 µm.

5.2 Fluorescence Microscopy Setup

For fluorescence excitation a 50 mW diode laser with an emission wavelength of λ=532 nm (VA-Serie; Roithner Laser Technik, Wien, Austria) was employed. The light was coupled into the light path by a dichroic mirror, 550 nm cut-off frequency (AHF Analysentechnik, Tübingen) and focused into the back focal plane aperture of the objective. The laser was used at full power and modulated by an external trigger signal to minimize bleaching of the fluorophores and obtain sharp images at the same time.

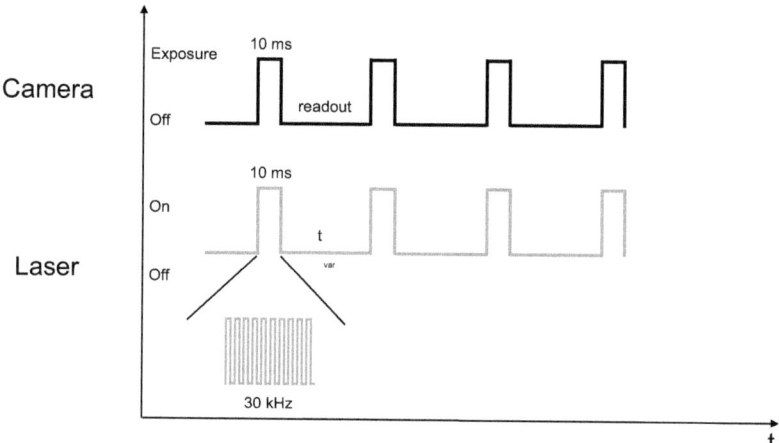

Figure 5.2: Exposure of the camera and excitation laser are synchronized. The camera records for 10 ms and only during this time the laser is active. Additionally, the laser is modulated with a 30 kHz frequency signal which allows the excited dye molecules to relax. Triplet excitation is minimized, which leads to reduced bleaching of the chromophores.

Fluorescence images were recorded on a peltier cooled CCD camera (1360 x 1036 Retiga EX; QImaging, Surrey, Canada). Using a binning of 2x2, exposure times of 10 ms were sufficient to record sharp images. The short exposure time reduced also the effect of blurring due to brownian motion of the filaments, which would be integrated at longer exposure times. Camera and laser were triggered by an external TTL signal from a data acquisition card (DAQPad-6229; National Instruments, Austin, TX). The frame rate of the camera could be adjusted by modulation of the dwell time between signals to up to 20 frames per second. But in order to reduce bleaching, typical frame rates were only 1.5 frames per second which was sufficient for most observations and additionally allowed experiment times of several minutes. The additional modulation of the laser at 30 kHz allowed the dye molecules to relax between each pulse to reduce bleaching by multiple excitation of metastable states. This was the fastest possible modulation frequency for the employed laser. The modulation scheme of the fluorescence laser and the camera exposure is depicted in figure 5.2.

The image was reflected into the camera by a dichroic mirror, transmitting above 593 nm (HC Beamsplitter BS593; AHF Analysentechnik, Tübingen). It was focused onto the chip by the same lens that focused the fluorescence laser into the back focal plane of the objective. The camera was protected against the green light of the laser, the transmitted light of the LED and the IR light of the trapping laser by a bandpass filter (BrightLine HC 572/27; AHF Analysentechnik, Tübingen) and an additional IR filter (KG3; Thorlabs, Newton, NJ). Together with the beamsplitters in the light path this allowed the recording of fluorescence images at good signal-to-noise ratio even at full power illumination by the LED of the brightfield imaging setup.

5.3 Brightfield Setup

Brightfield imaging was performed in this setup, using a small banded LED at 630 nm (Collimated Red LED, LEDC28; Thorlabs, Newton, NJ) as illumination source. This allowed for the spectral filtering of the brightfield illumination which is important when simultaneously imaging in fluorescence mode. Brightfield imaging was used in low magnification to observe the filling of the microfluidic channels in the upright microscope and in high magnification to perform particle tracking of trapped beads in the inverted microscope part of the setup.

5.3.1 High Magnification Inverted Microscope Setup

High magnification high-speed imaging was performed using the same objective as for the optical trapping. This guaranteed that the trapped objects were always in focus and facilitated imaging. The red LED was used to illuminate the sample in transmission mode and the light was recorded on a special high-speed camera (Phantom V7.2; Vision Research, Wayne, NJ) which allowed frame rates of more than 10000 fps. The beam splitters in the setup were chosen such that the red light at 630 nm was reflected only by the first beam splitter, a cold mirror, to couple in the IR trapping laser and transmitted by all others. The camera was protected against light of any unwanted wavelength by a band pass filter (BrightLine HC 628/40; AHF Analysentechnik, Tübingen) and an additional IR filter (KG3; Thorlabs, Newton,

NJ). The recording of the phantom camera was synchronized with the recording of the fluorescence camera via the data acquisition card and a labview software (LabVIEW v8.2; National Instruments, Austin, TX). Thus, it could be guaranteed that the capture of both high magnification cameras starts exactly at the same time which facilitated subsequent data analysis.

5.3.2 Low Magnification Upright Microscope Setup

The flow cells and the filling of the different channels during the experiment were visualized in the low magnification upright microscope part of the setup. To have an overview and to allow an effective control over the fluidic control a large field of view was required for this purpose, whereas resolution was only of minor importance. The illumination of the LED was coupled into the light path by a 50/50 beam splitter. Imaging was achieved by a low magnification objective (2.5 x Plan Neofluar; Zeiss, Oberkochen) and recorded on a basic CCD camera (AxioCam; Zeiss, Oberkochen). This objective also served as condenser for the illumination; during high-speed recording in the inverted setup it was therefore replaced by a 20 x objective (20 x/0.4 LWD, Nikon, Amstelveen, Netherlands) to collect more light into the field of view of the high-magnificatin setup. Figure 5.3 shows a compilation of the different images that could be simultaneously recorded in fluorescence, high magnification brightfield and low magnification brightfield imaging.

5.4 Setup Control

During the experiment, the different components of the setup had to be controlled partially mechanically and partially via electronic signals. We want to introduce now the different control and adjustment possibilities during the experiment and how they can be influenced by the experimenter. The upright microscope could be lifted in z-direction to adjust the focus using an implemented remote control. The same control unit would also move the bottom objective via stepper motors in x-, y- and z-direction to change the focus of the inverted microscope and to align the trapping laser. The sample stage was manually moved in x and y using a micrometer stage and additionally in nanometer precision in x-, y- and z-direction by a

5.5 Setup Control

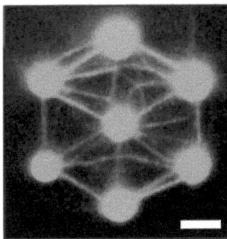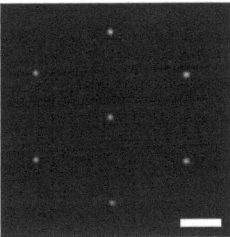

Figure 5.3: Example images that could be simultaneously obtained with the different cameras in the setup. On the left, a fluorescence image of phalloidin-TRITC stained actin filaments between an array of trapped beads. The beads fluoresce due to bound actin. The center panel shows the simultaneously recorded brightfield image of the bead array. Exposure time and illumination intensity are chosen very low to obtain perfectly Gauss shaped signals for the beads and low background to allow particle tracking. The right panel displays an image of an empty microchannel system imaged with the 2.5 x objective. The field of view contains the whole part of interest of the flow cell. Scale bars are 2 μm for the high magnification images and 200 μm for the low magnification image of the channels.

piezo driven stage (Physik Instrumente, Karlsruhe). The filling of the microfluidic channels was manually controlled by microliter syringes (10 μl♭ ILS Innovative Labor Systeme, Stützerbach) that were actuated by micrometer screws. In the case of pillar flow-cells, consisting only of a single channel, fluidic control was achieved by a motorized syringe pump. See section Microfluidic Flow Cells, 6.5, for details.

The power of the trapping laser, the LED illumination, the modulation of the fluorescence laser and the settings for fluorescence camera as well as the simultaneous triggering of the two cameras in the inverted microscope were controlled by a labview routine, that is shown in figure 5.4. Additionally, this routine performed real time image processing on the fluorescence image to provide a maximum range of contrast. The spatial light modulator and the piezeo stage were controlled by further labview programs. The phantom camera and the axiocam were controlled by the original programs of the manufacturer (phantom 640; Visionresearch and Axiovision; Zeiss)

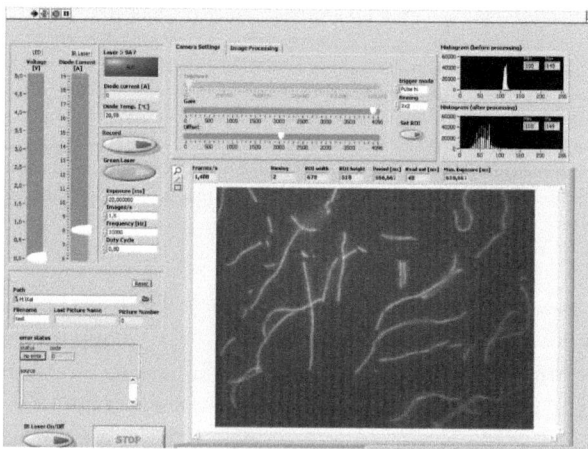

Figure 5.4: A screen shot of the software platform to control the setup: Intensity of the LED illumination and the IR trapping laser can be controlled by the virtual instruments. Exposure time of the fluorescence camera and duty cycle of the fluorescence laser are synchronized and can be varied. The live image of the fluorescence camera is displayed with enhanced contrast.

5.5 Tracking Procedures

Particle tracking for calibration of the optical traps and force and position measurements during the experiments were performed on high-magnification brightfield images using tracking routines developed by J.C. Crocker and D. Grier [195], originally written in IDL and transferred to MATLAB code by D. Blair and E. Dufresne. (Download: http://physics.georgetown.edu/matlab/): After spatially filtering and processing of the images to reduce background signals, a two dimensional Gaussian profile was fitted to the images. The center of such an Gaussian profile can be localized with sub-pixel accuracy. Since particles imaged in brightfield are represented by a Gaussian profile, this allowed the localization of particles within nanometer precision which is an order of magnitude better than the optical resolution of a light microscope. This procedure yielded a position versus time data set for each particle, which could be used for further analysis.

Chapter

6

Microfluidic Flow Cells

In this work, microfluidic flow cells were combined with an optical tweezers setup. This allowed the creation of complex multicomponent protein structures and the control over the chemical environment of the system at any time of the experiment. Microfluidic cells were a crucial compound for these experiments since a conventional exchange of solutions would easily lead to hydrodynamic forces which could either overcome the forces of the optical tweezers and therefore affect the experiment or even impact the protein structures and destroy them.

Two different types of flow cells were employed in this work: a multi channel flow cell for controlled exchange of environments during experiments with holographic optical tweezers and a single channel flow cell containing micropillar arrays [196], which acted as scaffolds for protein networks that were probed by single optical tweezers. From now on, the first one will be termed channel flow cell and the second one pillar flow cell.

The flow cells were produced by soft lithographic methods in polydimethylsiloxane (PDMS) [197, 198]. A negative mould was fabricated by photo lithography in a photosensitive resist. The structures were casted onto this mould and the flow cells assembled after peeling of the elastomer from the mold. The preparation of these two types of flow cells was widely similar. Thus, we will present them here in common and highlight the differences within the protocols.

6.1 Photolithography

In a first step, a photomask had to be produced to provide the structure of the mould. This was also achieved by photolithography using a positive photo resist (AZ-1505, Microchemicals, Ulm). All photolithographic steps were carried out under clean room conditions: A non-transparent chromium layer (120 nm) was deposited onto a glass slide (5x5 cm) by sputter coating (Sputter Coater MED020; Bal-Tec, Balzers, Liechtenstein) followed by spin coating of the resin onto the substrate (25 s, 3500 U/min). The structure was designed in AutoCAD software (Autodesk, San Rafael, CA) and written into the resist using a laser mask writer (DWL66; Heidelberg Instruments, Heidelberg). In a positive photo resist, as AZ-1505 is, the exposed structure will dissolve in the subsequent development process (1 min AZ351 Developer, 1:6 diluted in water). Thus, the chromium was unprotected at the exposed regions and could be dissolved in a chromium etching solution (Merck, 1 min, 1:3 diluted in water) uncovering the clear glass of the slide. The so fabricated mask could be used for the following exposure of the negative photo resist forming the mould after removal of the residual resin in acetone.

For pillar masks the resist was applied directly onto a glass slide, exposed and developed. On top of the developed structure, a chromium layer was sputtered and, subsequently, the resin was dissolved in acetone. Since only at the unexposed areas resin was left after the development process, it lifted off the chromium at this places when being solved by the acetone. Therefore, it was possible to obtain negative structures compared to the first procedure, which was more convenient when creating extended pillar fields.

The casting molds were produced in a negative epoxy resist (SU8; MicroChem Corp., Newton, MA) using a mask aligner (MJB3, SÜSS MicroTec, Garching). The resin was applied on silicon waver substrates (2" P/Bor 100; Si-Mat, Landsberg) by spin coating. The height of the obtained structures could be tuned by the choice of the resin and the parameters of the spin coating. The sample was soft baked on a hot plate to evaporate the solvent and exposed through the fabricated masks in the mask aligner. Following the exposure, a post exposure bake was performed to selectively cross-link the exposed portions of the film. The sample had to be cooled down slowly afterwards to prevent mechanical tension in the film. In the following development process, the not cross-linked resist dissolved and the exposed structures formed a solid and mechanically stable mould on the silicon waver. The technical parameters

Table 6.1: Photolithographic process of SU-8 mould fabrication

Process Step	Parameter	
	30 µm channel structures	15 µm pillar structures
Waver preparation	drying, 15 min 200°C	drying, 15 min 200°C
Application of resin	Spin coating, 2 ml SU8-25: 20 s at 500 U/min 30 s at 1500 U/min	Spin coating, 2 ml SU8-10: 10 s at 500 U/min 40 s at 2000 U/min
Soft bake	Hot plate: 3 min at 65°C 7 min at 95°C	2 min at 65°C 5 min at 95°C
Exposure	400 W HBO, (MJB3) 7 s	400 W HBO, (MJB3) 1.5-1.8 s
Post exposure bake	Hot plate: 1 min at 65°C 3 min at 95°C	1 min at 65°C 4 min at 95°C
Development	2 x 2 min in SU8 developer	3 min in SU8 developer

for the fabrication of the SU-8 structures are outlined in table 6.1.

Finally, the wavers were passivated to facilitate the later detachment of the siloxane when casting the elastomer structures. 1H,1H,2H,2H-Perfluorooctyltrichlorosilane (ABCR, Karlsruhe) was applied to the surface by chemical vapor deposition. For this purpose, the wavers were placed into an desiccator together with 100 µl of the silane. The desiccator was evacuated for 2 min and then kept closed for 2 hours. The evaporated silane reacted with Si-OH groups on the surface of the waver and air moisture to form a covalently bound layer. Subsequently, wavers were rinsed with ethanol and water to remove any unreacted silane.

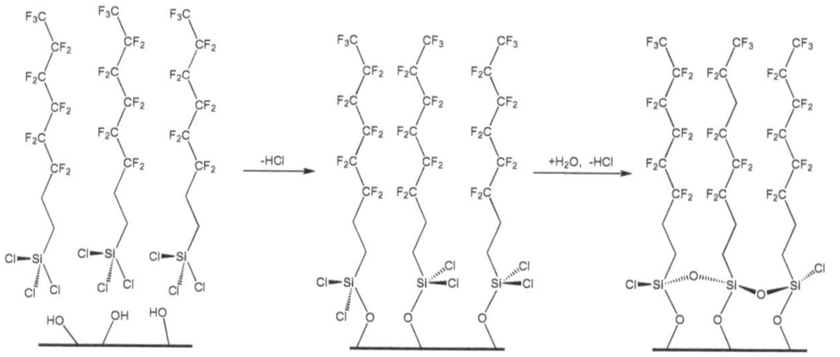

Figure 6.1: Reaction scheme of the passivation process of the silicon wavers with 1H,1H,2H,2H-Perfluorooctyltrichlorosilane. The chlorosilane reacts with surface bound OH-moieties under formation of HCl. In the presence of water the residual chloride groups can cross-link the silane to form a stable surface layer.

6.2 PDMS Casting

The structures were casted from the moulds in polydimethylsiloxane (PDMS). The employed PDMS (Sylgard 184; Dow Corning, Midland, MI) consists of about 250 monomers and is terminated by a vinyl moiety. This was coupled in the cross-linking process with added hydrosilanes via a platinum-catalyzed hydrosilylation. 10 ml of the silane were mixed with 1 ml of the cross-linking solution in a petri dish and evacuated for 2 hours in a desiccator to remove air from the elastomer which would affect the quality of the structures.

To create channel structures, 0.5 ml PDMS was casted onto the wavers and a cleaned coverslip (24x60 mm; Carl Roth, Karlsruhe) was pressed onto the structure. The mould together with the coverslip were cured over night at 65°C. Subsequently, the coverslips and the attached PDMS structure could be detached from the waver using razor blades.

To create pillar structures, a 5 mm thick layer of PDMS was casted onto the waver in a petri dish. The sample wad evacuated in the desiccator to remove air and subsequently cured over night at 65°C. The PDMS block could be peeled from the waver and cut in form with razor blades.

Wavers could be cleaned with ethanol and reused multiple times, but the passivation with perfluorosilane had to be repeated frequently.

6.3 Assembly of Flow Cells

The channel flow cell consisted of a PDMS-structured cover slip and a second coverslip bearing the tubing for the reactant supply. The top coverslip with the tubing was produced by drilling holes into a coverslip and attaching poly(ethylene) tubing (Portex, 0.28 mm inner diameter, 0.61 mm outer diameter, Smiths Medical, Watford, UK). The tubing was fixed using epoxy resin (UHU plus, schnellfest; UHU GmbH, Bühl) and the supernatant ends of the tubing at the bottom side of the coverslip were removed with a razor blade.

The pillar flow cells were composed of a coverslip, structured with a single channel in PDMS, wide enough to incorporate the pillar field as bottom part and a PDMS block with the pillars and tubing as the top part. The tubing was mounted to the PDMS block into holes, produced by biopsy punches (0.75 mm diameter, Harris Uni-Core; Ted Pella Inc., Redding, CA) [199]. The tubing was fixed on the polymer by silicon rubber (Twinsil; Picodent GmbH, Wipperfürth).

To mount the two parts of the respective flow cells, their surfaces had to be rendered adhesive to each other. The top of the channel flow cell, being a glass surface was immersed for 10 min into an alkaline solution (Extran, Merck, Darmstadt) to slightly etch the surface. The PDMS covered surfaces, *i.e.* the channels and the pillar field were activated by oxygen plasma (30 s, 150 W, 0.5 mbar; 100-E; TePla AG, Wettenberg). The plasma oxidized the methyl groups of the siloxane, leaving $Si-O^-$ moieties at the surface which could form stable bonds to the other activated surface [200–202]. The pillar flow cells were additionally sealed at the seam between PDMS and the glass by the same silicon rubber, used also for the fixation of the tubing. Both flow cell designs and the scheme of the assembly are shown in figure 6.2

6.4 Cleaning of Flow Cells

The pillar flow cells could only be used once, since proteins and reactants adhering to the pillars cannot be washed off after experiment. Channel flow cells were rinsed using the following procedure and be reused several times:

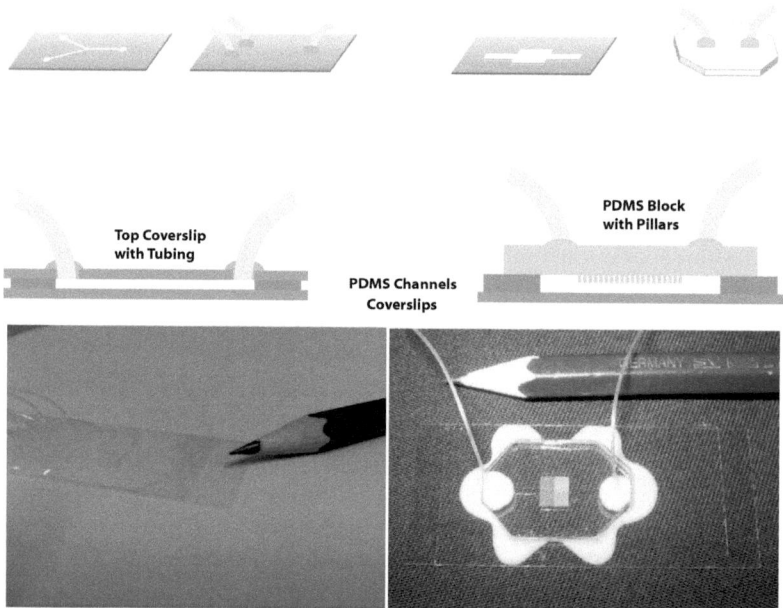

Figure 6.2: Comparison of the two flow cell designs presented. The left panel shows a channel flow cell consisting of two coverslips. In- and outlets are placed on the left side of the flow cell, fixed with epoxy resin. The right panel displays a pillar flow cell. The PDMS block, attached to the bottom cover slip, is cut in a hexagonal form and sealed as well as the tubing by silicone rubber. Right picture courtesy of Timo Maier.

- Rinsing with MilliQ water, 10 ml
- Rinsing with Extran solution, 1:10 in water, 5 ml
- Rinsing with MilliQ water, 10 ml
- Rinsing with Ethanol, 5 ml
- Rinsing with MilliQ water, 10 ml
- Drying at 65°C for 10 hours

The success of the cleaning procedure was controlled by filling of the flow cell with water; a clean flow cell showed a high hydrophobicity and the wetted area was thus minimized, whereas a contaminated flow cells showed increased wetting.

6.5 Flow Cell Design and Scheme of Usage

The pillar flow cells consist only of a single channel which has an inlet and an outlet on two sides of the pillar field. Reactants were therefore pumped serially into the flow cell and to the pillar field by an automated syringe pump (Harvard Pump 11 Plus; Harvard Apparatus, Holliston, MA) at flow velocities in the order of 0.5-5 µl/min. To add reactants, the end of the inlet tube was placed into a vial and the outlet was connected to the syringe pump. Exchange of solutions could be achieved by just changing the vial. Hydrodynamic forces of the flow during fluidic exchange in this system were too high to allow a simultaneous application of optical tweezers which could only be used after stopping of the flow.

The channel flow cell was designed to allow the combination of holographic optical tweezers and media exchange. The fluidic system contains five channels of 100 µm width for feed-in of reactants and one channel of 350 µm width where the measurements were performed, the microreactor (see also figure 5.3 right side and figure 9.2 for details). This channel was wider than the others to balance pressure peaks during fluidic events. In the experiment it was blocked by an air bubble and due to high friction of air-water-interfaces in PDMS channels prevented any unwanted influx of liquids during the experiment. Additionally, one 550 µm wide channel, was connected to the outside of the cell to compensate pressure differences in the system . Micro-syringes filled with the chemically inert perfluorodecailine (Sigma Aldrich,

St. Louis, MO), actuated with micrometer screws, were used to add and retract solutions. Liquid amounts in the order of nanoliters could be controlled this way without causing significant flow. To fill the flow cell, the reactants were injected into the respective tube followed by connection of the tube to the syringes. By actuating the syringes, it was therefore possible to push the solutions via the decaline into the channels. Adding solution into one channel did not affect the other channels and it was possible to mix different solutions without loosing trapped particles, as shown in the experiments chapters (9.1, 10.1 and 11.2).

6.6 Surface Modification of Micropillars

To provide adhesive surfaces for actin deposition on the pillar tops, the substrate had to be chemically modified. As interaction partner, NEMHMM, a modified myosin molecule was chosen. NEMHMM (N-Ethyl-Maleimide-modified Heavy MeroMyosin) is a myosin fragment that was enzymatically cleaved to provide only the head and neck region of the molecule. Its ATPase activity is inactivated by maleimide addition. The resulting molecule has an increased solubility and stability compared to native myosin and keeps its actin binding ability even in the presence of ATP. The molecule binds to the pillar heads by hydrophobic interaction and provides binding sites for actin filaments. Coating of the pillars had to be performed directly before the experiment to ensure functionality of the protein. The pillar flow cell was rinsed with a solution of 50% ethanol in water to guarantee complete wetting of the surface. After rinsing with D-buffer (see chapter 7.3), the pillars were incubated for 30 minutes with NEMHMM at a concentration of $5\,\mu M$. After rinsing again with D-buffer, the actin could be added and adhered to the pillars.

Chapter

7

Proteins and Buffer Solutions

Proteins in solution are very sensitive to degradation by bacterial proteases and oxidative processes [203–205]. In the case of actin, this leads to significantly different polymerization and cross-linking behavior and therefore to misleading viscoelastic properties [206, 207]. To obtain comparable results, actin was always freshly polymerized and diluted. Diluted filamentous actin (F-actin) could be only used for one day. Polymerized actin in the original concentration of 5 µM was stored on ice for up to 14 days, but had to be controlled before each experiment for bundle formation and filament fragmentation. Unpolymerized, globular actin (G-actin) can be kept in solution only for several days.

A typical actin preparation yielded up to 100 mg of protein, but for an experiment less than 200 ng were needed. Therefore, the preparation was designed, to be divided multiple times and stored in aliquots at -80°C to provide fresh samples at any time.

7.1 Actin Isolation and Purification

Monomeric actin (G-actin) was prepared from rabbit skeletal muscle following the standard method of Pardee and Spudich [208]. To remove residual cross-linking and capping proteins, it was purified by an additional gel column chromatography (Sephacryl S-300) step as described by MacLean-Fletcher and Pollard [209]. To guarantee the absence of any residual tropomyosin, which could alter cross-linking properties of the actin, only the late fractions of the molecule peak from the column chromatography were taken for further use. The G-actin was divided into aliquots of 3 mg and lyophilized subsequently. The freeze dried aliquots could be stored at

-80°C for more than 12 months.

7.2 Actin Storage and Dialysis

Prior to the polymerization, the freeze dried samples had to be diluted again and the elution buffer from the chromatography had to be removed. The aliquot of 3 mg was dissolved in 2 ml MilliQ water and dialyzed against G-buffer.

Table 7.1: G-buffer, pH: 8.0 (MW: molecular weight, c: concentration, m: content in 1000 ml buffer)

Substance	MW in [g/mol]	c in [mmol/l]	m [mg]
TRIS	121	2.0	242
$CaCl_2 \cdot 2H_2O$	147	0.2	19
DTT	154	0.2	31
NaN_3 (20% solution)	65	3.0	1 ml
Na_2ATP	551	0.2	110

The G-buffer contained DTT (dithiothreitol) to prevent oxidation of actin and sodium azide as a biozide in order to preserve the actin. Dialysis was done at 4°C using dialysis casettes (Slid-A-Lyzer, MWCO 7000; Perbio, Bonn) for 20 hours. The solution was centrifuged at 100.000 g for 2 hours (Sorvall Discovery M 120 SE; Thermo Scientific, Waltham, MA) to remove degraded protein and bacteria.
Actin concentration was determined by UV-Vis spectroscopy (Nanodrop 1000; Thermo Scientific, Waltham, MA) by measuring absorbance at $\lambda = 290$ nm and assuming an extinction coefficient e^{290} of 0.63 ml/mg. To prevent the formation of ice crystals during the following freezing process 20% ethylene glycol were added. Aliquots of 0.5 nmol actin were taken and frozen by dropping into liquid nitrogen. Due to the fast freezing and the glycol, no crystallites could form and thus the protein was not harmed. The frozen droplets were collected in PCR vials and stored at -80°C .

7.3 Actin Polymerization and Staining

Actin polymerization was accomplished by a polymerization buffer, the F-buffer. It was initiated by the high salt concentration and the ATP in the buffer. The actin filaments were stained and stabilized by phalloidin-TRITC, a fluorescently labeled

7.4 Actin Polymerization and Staining

derivative of the toxin of the death cap mushroom (*Amanita phalloides*). F-buffer was prepared as a 10x stock and stored in 50 µl aliquots at -20°C since the ATP would hydrolyze in solution. To initiate polymerization, F-buffer was added to one

Table 7.2: F-buffer, pH: 7.4 (MW: molecular weight, c: concentration, m: content in 100 ml 10x buffer)

Substance	MW in [g/mol]	c in [mmol/l]	m [mg]
TRIS	121	2.0	240
$MgCl_2 \cdot 6H_2O$	203	2.0	410
KCl	75	100.0	7500
$CaCl_2 \cdot 2H_2O$	147	0.2	30
DTT	154	0.2	30
MgATP	507	0.5	250

aliquot of actin to give a final volume of 95 µl. The solution was mixed gently and kept at 4°C to yield a small number of nucleation cores. After 2 hours, 5 µl of phalloidin-TRITC solution (0.1 mM in methanol, Sigma-Aldrich) were added. This resulted in a final concentration of 5 µM actin. After 24 hours the polymerization was completed and the filaments could be stored at 4°C for up to 14 days.

Prior to the experiment, the actin had to be further diluted in dilution buffer (D-buffer), which was also prepared as a 10x stock solution and kept in 2 ml aliquots at -20°C.

Table 7.3: D-buffer, pH: 7.4 (MW: molecular weight, c: concentration, m: content in 100 ml 10x buffer)

Substance	MW in [g/mol]	c in [mmol/l]	m [g]
Imidazol	69	25.0	1.70
EGTA	380	1.0	0.38
$MgCl_2 \cdot 6H_2O$	203	4:0	0.81
KCl	75	25.0	1.87

Before experiments, the actin had to be controlled for filament length and unspecific bundling. Therefore, a dilution of 1:600 in D-buffer was analyzed by confocal laser scanning microscopy (LSM Pascal 5; Zeiss). Average filament length should exceed 10 µm and all filaments should be of similar brightness to exclude the presence of bundles. See figure 7.1 for a typical confocal fluorescence micrograph as taken before each experiment as control.

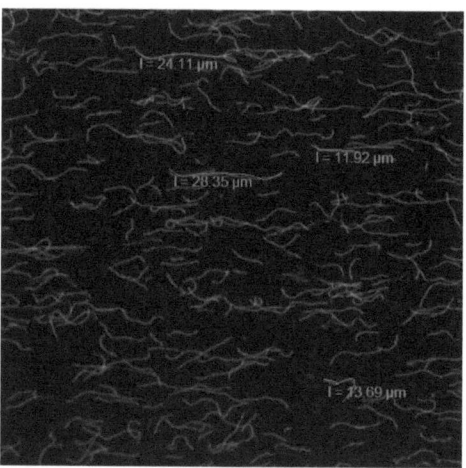

Figure 7.1: The figure shows an example fluorescence micrograph of actin filaments attached to the glass surface to control their average length and the absence of any unspecific bundles.

7.4 α-Actinin

α-actinin was purchased from Sigma-Aldrich as ammonium sulfate suspension (4 mg/ml). In this form it is very stable against degradation but it is not accessible for crosslinking. Therefore, the protein had to be dialyzed prior to the experiments. 250 µl of the suspension was centrifuged for 30 min at 50.000 g to collect the precipitate. The supernatant was removed and the α-actinin was resuspended in 250 µl D-buffer containing 15 mM DTT. DTT is necessary to support the solution of the α-actinin and cleave intermolecular disulfide bonds. The solution was mixed for 2 hours at 4°C and dialyzed in mini dialysis units (Mini Slide-A-Lyzer, Pierce) against D-buffer containing 0.2 mM DTT and 3 mM NaN_3.

Concentration of α-actinin is determined by UV-Vis spectroscopy, using an absorption coefficient e^{280} of 1.238 ml/mg ($\hat{=}$ 127365 $M^{-1}cm^{-1}$).

7.5 Adhesive Microparticles

Commercially available microparticles are usually composed of latex or polystyrene. To prevent these from clustering they bear negative surface charges (sulfate moieties) and are suspended in surfactant containing solutions. This renders them inadhesive to the negatively charged actin filaments. To coat the beads with positive surface charges, poly-L-lysine was implemented. Due to entropic and electrostatic effects the lysine binds to the surfaces of the particles and the free amino groups provide positive charges when protonated and therefore possible adhesion sites for actin.

50 µl bead suspension (2 µm Polybead polystyrene beads, 2 % suspension; Polysciences, Eppelheim) were washed three times with MilliQ water by centrifugation and resuspension (Biofuge Fraesco; Heraeus Centrifuges,Buckinghamshire, England). 10 µl of the washed bead suspension were added to 90 µl solution containing 5 mg/ml Poly-L-Lysine (MW 15.000-30.000; Sigma Aldrich). Beads were kept at 4°C at continuos agitation to prevent clustering. Directly before experiments, the beads were washed again three times with MilliQ water by centrifugation and resuspension in water or buffer solution according to requirements.

Chapter

8

Plasmodium Sporozoite Experiments

The part of this work involving *Plasmodium* sporozoites was executed in collaboration with the group of Friedrich Frischknecht at the department of parasitology at the University of Heidelberg. *Plasmodium* culturing and preparation was performed at this group. For reasons of completeness, a short subsumption of different working steps to prepare sporozoite samples is given.

8.1 Preparation of *Plasmodium berghei* sporozoites

As sporozoites do not replicate by simple cell division they cannot be kept in cell culture environments. Instead, the complete live cycle of *Plasmodium* has to be provided, including vertebrate and insect hosts. *Plasmodium berghei* infected mice were exposed to *Anopheles stephensi* mosquitoes. After 16 to 25 days maturation inside the mosquito, the salivary glands were dissected and sporozoites extracted. Sporozoites were kept in RPMI buffer (RPMI 1640, no phenolphthalein, glucose provided; PAA Laboratories, Cölbe) at 4°C for up to 5 hours and could be used without significant changes in viability and behavior during that time. GFP (Green Fluorescent Protein) transfected parasites were used to allow fluorescence imaging and to facilitate dissection. Knock out parasites known to fail to invade mosquitoes salivary glands (e.g. TRAP-KO) were extracted from the insects' hemolymph. Genetically modified sporozoites were kindly provided by the group of Kai Matuschewski, department of parasitology at the University of Heidelberg. Photographs of fluorescent sporozoites inside the mosquito and the salivary glands as well as an isolated sporozoite are shown in figure 8.1

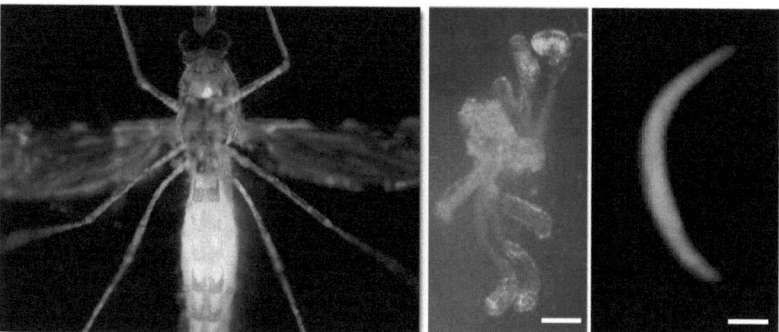

Figure 8.1: Left: GFP transfected *Plasmodium* sporozoites inside the mosquito. Accumulations inside the gut and at the salivary glands region are observable. Middle: Magnification of extracted salivary glands of infected mosquito. Right: Single sporozoite in fluorescence microscopy Scale bars are 100 µm (middle) and 2 µm (right). Courtesy of Stephan Hegge.

8.2 Optical Tweezers Experiments with Sporozoites

Optical tweezers experiments with sporozoites were performed in simple home made microscopy chambers consisting of 2 coverslips, connected by double sided sticky tape as a spacer. Into the tape, a chamber of about 5 x 5 mm width was cut as reservoir. Freshly dissected sporozoites were kept on ice and used for up to 5 hours after dissection. However fresh samples were taken after 60 minutes from the chilled stock since sporozoites undergo transformation at room temperature after 120 minutes. BSA (Roth, Karlsruhe) was added to the sample solution to a final concentration of 3% weight to enhance adhesion [171]. SYTOX Orange (Invitrogen, Karlsruhe) was employed at a concentration of 5 µM as nuclear fluorescence staining. This allowed the determination of the orientation of the sporozoites and a control for sporozoite viability. In actin deficient experiments, cytochalasin D (Merck Biosciences, Nottingham, UK) was used as actin depolymerizing drug at a concentration of 2 µM. Typical laser power for adhesion forming experiments was 100 mW in the sample plane. In experiments to detach sporozoites, trapping powers up to 450 mW were used. Sporozoites were trapped and manipulated in the xy-plane using the micrometer table and in z-direction using the piezo stage, controlled by labview routines.

Part III

Experiments and Results

Chapter

9

Two-Dimensional Actin Networks

The cell cortex is only a few hundred nanometers thick but extends over the whole cell membrane, which can be tens of microns [210, 211]. Hence, it can be considered as a quasi two-dimensional network. Its appearance is controlled by reversible polymerization of actin fibers, actin cross-linking proteins and motor proteins that regulate the viscoelastic and dynamic properties of this structure. Cells actively control the network to adjust their polarity, their morphology, their motility and cellular processes like adhesion, division or exo- and endocytosis [34, 106, 109, 151, 212, 213]. From the physical point of view, the actin cortex is a sterically entangled and partially cross-linked solution. The persistence length l_p of actin is in the same order of magnitude as the contour length l_c as well as the mesh size ξ of the network [121, 214, 215]. The mechanical behavior of such networks significantly differs from macroscopic models and the unusual viscoelastic properties have drawn attention in recent years [216–218].

It would be desirable to investigate how cells can change the structure of their cortical actin to generate forces and adapt to their environment. To reduce the complexity of the system, biomimetic approaches are most promising to get insight into the chemomechanical coupling of the actin cortex. Until now, most studies of biomimetic actin networks only cover three-dimensional systems [115, 219–221]. Very few studies addressed the question of networks confined to flat geometries where the thickness is small compared to persistence length, contour length and mesh size [196, 222, 223] and none of them could provide micromechanical measurements of the observed systems.

Here, we want to present a method to construct entangled actin networks confined to a two-dimensional geometry. Using a combination of HOT, fluorescence microscopy and microfluidics a quasi two-dimensional network of actin filaments in between an array of optically trapped micro beads was created, allowing quantitative force probing in the piconewton regime. Thus, the time-dependent increase of mechanical tension in the protein network during bundling via magnesium ions could be detected. Figure 9.1 shows a model of the hexagonal bead array with attached actin filaments inside the microfluidic flow cell. To implement the holographic optical tweezers into the microfluidics, special flow cells and a flow cell operating methodology had to be developed. To have a complete control over the chemical microenvironment, flow cells with multiple parallel channels that would merge inside the flow cell were developed. An operating scheme based on the controlled addition and subsequent retraction of liquids in the different channels via attached microsyringes provided the required flexibility and operationability.

9.1 Experimental Procedure to Create and Probe Two-Dimensional Actin Networks

A flow cell containing seven different channels was used to create the filament network and measure the forces during the subsequent cross-linking of the actin structure. The system was designed to pass several different challenges:

- Exchanging solutions in an adequate timescale without generating forces, which could exceed the trapping strength of the holographic optical tweezers

- Maintaining no flow conditions throughout the force measurements so as to guarantee the desired force resolution

- Designing a device with high transparency to allow fluorescence imaging at the same time as high-speed brightfield imaging

- Minimizing the overall consumption of solutions for the experiment to allow the conduction of experiments with substances and proteins that are available only in small amounts

The experiment and the flow cells providing these required features were designed as follows: To provide a reservoir of free beads for trapping, beads had to

Experimental Procedure to Create and Probe Two-Dimensional Actin Networks

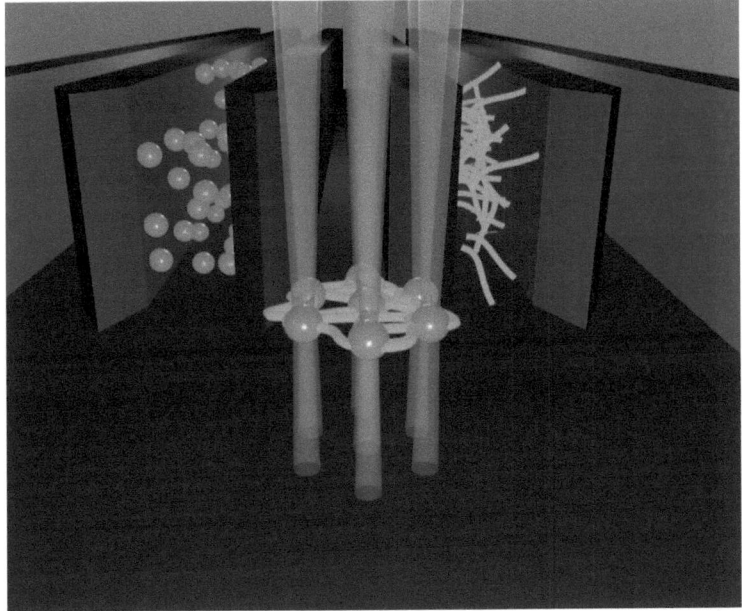

Figure 9.1: The figure shows a schematic representation of the two-dimensional actin network confined by seven beads trapped in holographic optical tweezers. The system is in a microfluidic environment as used in this work. In the background, an actin filled channel and a channel with bead suspension are displayed.

be suspended in water. In buffer solutions particles would adhere to the channel walls due to van der Waals forces, which scale with the ionic strength of solutions. Beads were flown in and trapped by the optical tweezers. After transferring the beads into buffer solution, actin filaments were added and attached to the beads and subsequently the filaments were cross-linked. However, during this step it was important to provide no-flow conditions to allow high-resolution force measurements during the resulting contraction of the network.

In the following we will describe the filling scheme of the different channels and how the liquids were operated to perform the measurement. See also figure 9.2.

The fluidic system contained five channels of 100 µm width for feed-in of reactants and one channel of 350 µm width where the measurements are done, the designated microreactor. This channel was wider than the others to balance pressure peaks

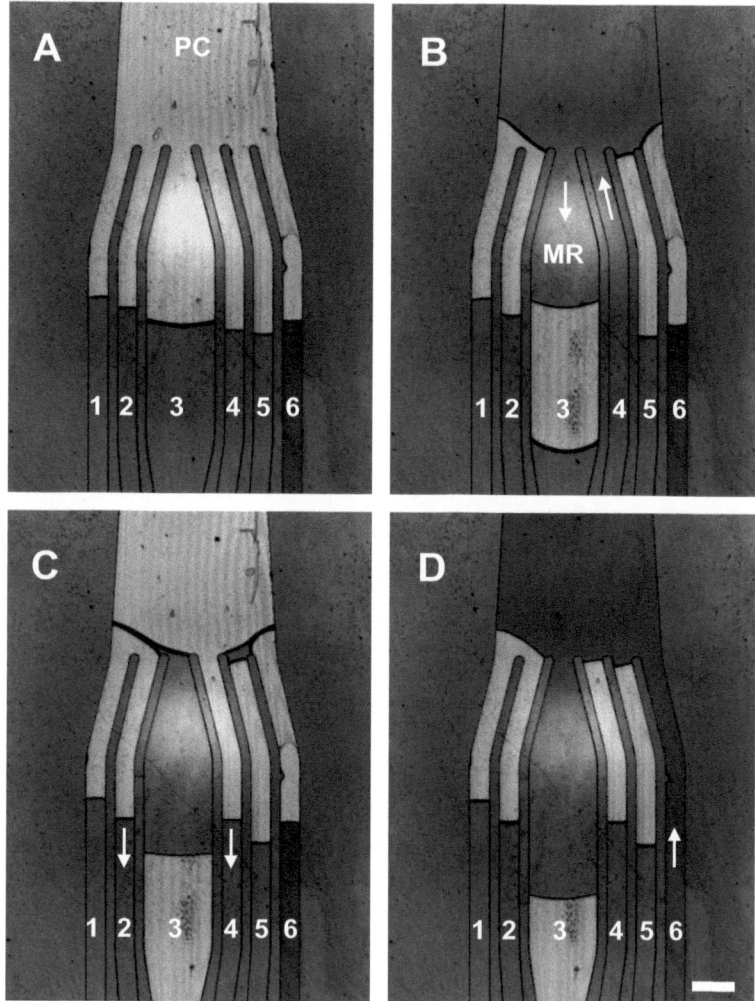

Figure 9.2: The figure shows an example operating procedure for the multi channel flow cell. Liquid channels appear dark grey, air has a white appearance. In the beginning (**A**), all channels are partially filled. In **B**, channel 4 is added and the microreactor filled. The air bubble in the lower part of channel 3 is clearly visible. In **C**, the liquid is removed into channel 2, whilst the reaction solution stays in the microreactor. In **D**, a second solution is added via channel 6, it can diffuse now into channel 3. Scale bar is 200 µm.

during fluidic events. During the experiment, it was blocked by an air bubble. Due to high friction of air-water-interfaces in PDMS channels, any unwanted influx of liquids during the experiment was prevented. Additionally, one channel, which was 550 μm wide, was connected to the outside of the cell to compensate pressure differences in the system. Micro-syringes actuated with micrometer screws were used to add and retract solutions. Liquid amounts in the order of nanoliters could be controlled this way without causing any significant unwanted flow. A sequence of different filling steps is shown in figure 9.2:
The loading of the different channels is listed in table 9.1. In the beginning, all channels except the pressure compensation channel (PC) were filled partially, leaving air in the last 500 μm of the single channels. Liquids in the channels appear dark grey; empty channels have a lighter appearance. Microchannels were filled as follows: Channel 1 contained the cross-linking solution of 100 mM $MgCl_2$ in D-buffer. Channel 2, filled with water, was employed as a drain to remove used solutions from the flow cell during the experiment. Channel 3 was the designated microreactor. It was filled with water up to 500 μm before the connection point of all channels. Channel 4 was filled with D-buffer. Channel 5 contained the actin filaments, diluted in D-buffer to a final concentration of 200 nM. Channel 6 contained the suspension of functionalized microbeads at a concentration of 0.13% to be trapped in the hexagonal HOT array and serve as scaffold and force sensors for the protein network. See chapter 7 for details on proteins, beads and buffers

To fill the microreactor (channel 3) with D-buffer, channel 4 was actuated and pressed into the PC channel. Subsequently, the solution was pulled into the microreactor, leaving an air bubble between the buffer and the original medium in channel 3 (figure 9.2 B). This measure prevents unwanted mixing and mechanically stabilizes the solution in the microreactor. Excessive liquid was retracted again, using the drain, channel 2 (figure 9.2 C). The amount of solution in the reactor was about 15 nL. In total, less than 50 nL were needed for one filling cycle. In the next cycle bead suspension was added, actuating channel 6 (figure 9.2 D). Seven beads were trapped in a hexagonal HOT pattern and then transferred to the microreactor by moving the micrometer stage of the microscope. The bead suspension was removed, leaving air in the pressure compensation channel. The beads remained optically trapped inside the microreactor, now exposed to the chemical environment

Table 9.1: Filling scheme for the different flow channels

Channel	Content	Purpose
1	100 mM MgCl$_2$	Cross-linking solution
2	Water	Drain to remove used solutions
3	Air/Water	Microreaction chamber
4	D-buffer	Environment for the experiment
5	Actin in D-buffer	Filaments to be collected between trapped beads
6	Beads in Water	Handles for force measurements

of D-buffer. Then, actin filaments were added using channel 5 and the beads were moved out of the microreactor into the actin solution, in order to collect actin. Multiple filaments attached to and stretched between the beads, as visualized using fluorescence microscopy (see figure 9.3, first image). Once a dense filament network was assembled, it was transferred back into the microreactor and the actin solution was removed. Finally the MgCl$_2$-solution was added via channel 1 and connected to the microreactor. The magnesium ions diffused into the microreactor where they started to immediately cross-link the filaments into an entangled network (see figure 9.3, third image). As the actin was cross-linked, contractile forces were exerted throughout the network and measured at the seven force sensors (calibrated optical traps) embedded in the network. This was done by monitoring the beadsÕ displacement from their equilibrium positions with high-speed brightfield microscopy (see figure 9.5).

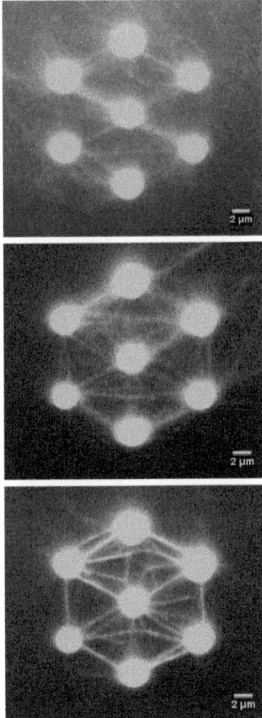

Figure 9.3: The figure shows a two-dimensional network of actin fibers between optically trapped microspheres. In the first image, the trapped pattern is brought into contact with the actin solution and filaments attach to the beads. The actin in solution is visible as bright background. In the second image, the network is transferred into the microreactor; no filaments are in the background and between the beads an entangled but not yet cross-linked network is formed. After addition of the cross-linker (third image), thick bundles form between the beads, hardly any free single filaments are visible anymore in the network.

9.1.1 Calibration of Trap Pattern

As was shown in chapter 2.5 multiple techniques to measure the stiffness of optical traps have been described in literature. Microfluidic systems are especially susceptible for noise due to fluctuations in the medium. Already minimal temperature differences in the channels can lead to unwanted flow and thus external forces on the optical traps. Therefore, calibration methods that can be influenced by such external factors are not suitable for calibration of the traps in the microfluidic system. Power spectral analysis of the beads motion inside the optical traps was chosen as the best method leading to reliable results for the trapping stiffnesses. Until now, this method has only been used for single traps. Suitable routines for the camera based calibration of multiple holographic optical traps have been developed for this work. To test the reliability of the calibration it was compared to standard calibration methods based on Boltzmann statistics.

Calibration of the seven optical tweezers was accomplished by analysis of the confined thermal motion of the trapped microspheres [224]. Assuming particle displacements occur within the linear Hooke's regime of the trap, the force F exerted on the microsphere is given by $F = -\kappa \cdot x$, where κ is the stiffness and x is the displacement with respect to the center of the optical trap. Two different calibration methods were used: Boltzmann statistics and power spectrum analysis. See sections 2.5.2 and 2.5.4 for details.

In the first case, the probability distribution of the displacements

$$P(x) \propto \exp\left(-\frac{1}{2}\frac{\kappa x^2}{k_B T}\right) \tag{9.1}$$

was calibrated, where k_B is the Boltzmann constant $(1.38 \cdot 10^{-23} J K^{-1})$ and T is the absolute temperature. By fitting a Gaussian to the experimental data, the stiffness κ was extracted. The second technique is based on the computation of a frequency (f) dependent power spectral density (PSD) $S(f)$. The theoretical PSD $S(f)$ is given by a one-sided Lorentzian

$$S(f) = \frac{D}{\pi^2(f^2 + f_c^2)}, \tag{9.2}$$

where D is the particlens diffusion constant, $f_c = \kappa/2\pi\gamma$ is the corner frequency and γ is the drag coefficient of the microsphere. An experimental PSD was computed

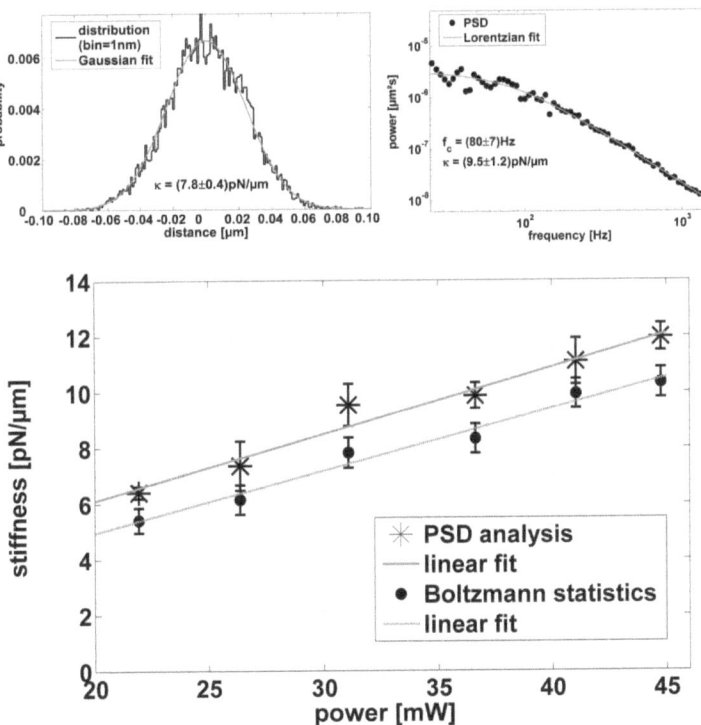

Figure 9.4: Example of calibration for the optical trap in the middle of the hexagon as obtained by Boltzmann statistics and power spectra analysis respectively. The small insets show calibration result for one exemplary laser strength and the big graph is a comparison of the results for the two methods presented. For each of six different laser intensities 12000 frames with a frame rate of 3 kHz were recorded. On the left a Gaussian fit for the probability distribution according to eq.: 9.1. Right: Power Spectra Density (PSD) analysis of the data set. Stiffness is obtained by eq.: 9.2, whereas a Lorentzian was fitted between 25 and 600 Hz. Brownian motion statistics systematically underestimates the real trapping force due to low frequency noise of the system (all errors are related to 68 % of confidence interval).

from the position data set by calculating the square modulus of the complex Fourier transform [225]. Subsequently, a block of n_b consecutive data points was replaced by a single new data point to obtain a smaller data set with less noise. The Lorentzian was fitted with $n_b = 50$ according to eq.: 9.2 . For the double logarithmic plot (see figure 9.4), increasing block size n_b was chosen to obtain equidistant data points. Therefore, the confidence level increases with the frequency.

Both methods require high-resolution position versus time information for each particle. The position data sets were generated by image processing based particle tracking algorithms. After video acquisition, the images of the trapped microspheres were tracked simultaneously with subpixel resolution. Calibration by Brownian statistics systematically underestimates the spring constants of the traps due to low frequency noise in the system, i.e flow channels, microscope setup and environment. In the case of calibration via a power spectra analysis, the fitting range can be chosen that way to exclude such sources of noise, resulting in more reliable values.

9.2 Contractile Forces during Cross-Linking of Actin Network

To measure the build up of mechanical tension in the network, additionally to the fluorescence imaging, a high-speed camera for the brightfield imaging of trapped beads was employed. Thus, the movement of the beads could be followed with approximately 5 nm precision and a temporal resolution in the millisecond range. Since all beads were in calibrated optical traps, their movement was directly related to the forces acting on each bead. A comparison for the positions of the beads in the optical traps before and after cross-linking is shown in figure 9.5.

To visualize the contractile behavior of this forces, the projection of the displacement of the particles along the diagonal axes of the hexagon was plotted. Figure 9.6 demonstrates as the forces acting on one diagonal of the hexagon (beads 2, 4 and 6, see left part of the figure). The direction of the forces in the right graph is related to the shown projection vector. Positive force values indicate forces along this projection vector in the direction indicated by the arrowhead, negative force values towards the other end. The force curves start (time: 0 s) when the actin filament network was established and separated in the microreactor channel. After 20 s the

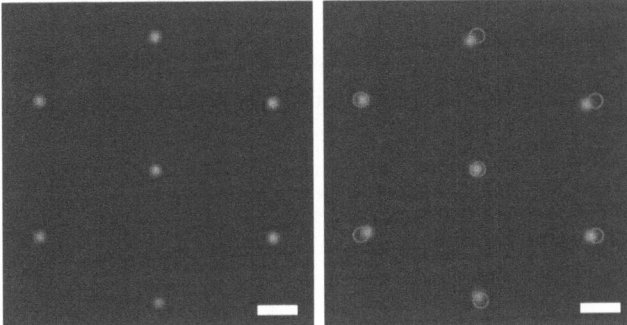

Figure 9.5: This figure shows the brightfield image of the trapped beads during the cross-linking experiment, as obtained with the high speed camera. The picture on the left side shows the particles before cross-linking. The beads are in the equilibrium position of the holographic traps, no deflections and thus no forces are detectable. The picture on the right side shows the particles after addition of cross-linker and contraction of the network. The beads are displaced from their equilibrium positions by the cross-linking forces. For comparison the positions of the beads before the contraction are marked by red circles. From the calibrated spring constants of the traps it is possible to calculate the force acting on each bead for each frame of the experiment. Scale bars are $2\,\mu m$.

channel with the magnesium ions was connected to the microreactor channel (black arrow on the force curve). The force curves show only a small reaction of the force sensor, demonstrating that the exchange of solutions could be achieved, causing forces below 1 pN on the trapped spheres. Now the magnesium ions could diffuse into the microreactor channel. The force curves of the beads of the show that the outer beads were pulled towards the center of the network during cross-linking (data shown for beads 2, 4 and 6). The maximum force that could be measured on the outer beads after the complete cross-linking of the network was 2-3 pN (time: 120-140 s). Almost no force acted on the center of the network, which is demonstrated by the force curve of the center bead (bead 4). As a comparison, the actin filaments have a stretching modulus of about 1.7 pN/nm for a 20 µm long filament and are reported to withstand forces up to about 200 pN [226]. Therefore, the elongations caused by the forces applied in our optical traps are below the detection limit and can be neglected in our measured data. Even though the piconewton forces acting on

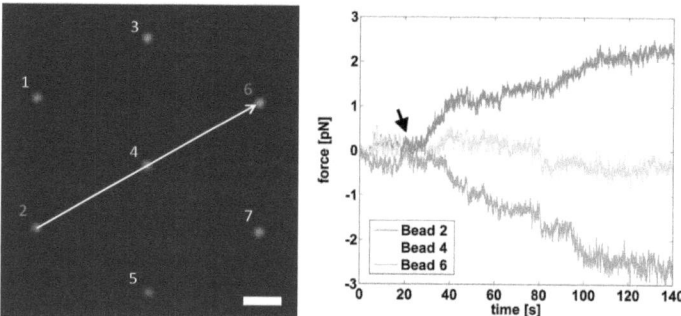

Figure 9.6: Contractile forces build up after connection of the MgCl$_2$-Channel with the microreactor (black arrow at t=20s). To visualize the directionality of the forces, the projection along the hexagons diagonal is depicted, as shown in the left image. Forces into the direction of the arrow are plotted as positive values. The graph shows only example data for three of the seven beads (2,4, and 6). The vector was obtained by connecting the centers of the respective beads before the cross-linking.

the beads seem to be relatively low, they are for instance in the same regime as forces of single molecular motors [71]. The experiment demonstrates the bundling of actin filaments in a quasi two-dimensional geometry, only due to counterion condensation caused by magnesium ions as Òcross-linkersÓ. In contrast to other cross-linkers of actin, magnesium ions are present almost everywhere in the cell, though at about some micromolar concentration, the charge shielding induced bundling may play an important role inside cells [227].

Furthermore it is possible to measure the total energy stored in all optical traps during the experiment:

$$E = \sum_{i=1}^{7} 1/2 k_i x^2. \qquad (9.3)$$

This was done as a function of the experiment time in figure 9.7. The energy reaches a plateau at about $600\,k_B T$, this energy is comparable to the energy of about 60 hydrogen bonds and gives an estimate of the number of molecules involved in the contraction.

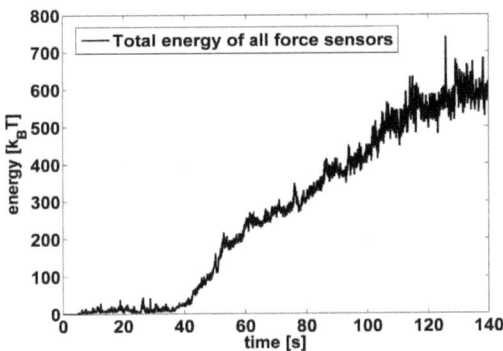

Figure 9.7: Total energy stored in the optical traps during cross-linking of the hexagon in units of $k_B T$.

9.3 Conclusions on Two-Dimensional Actin Networks

The combination of the HOT technology, fluorescence microscopy and a multichannel, stop-flow microfluidic device represents a novel, multipoint force sensor in the piconewton regime for constructing and probing extended, complex micro systems. We presented, the arrangement of multiple objects as force sensors while actively transforming the chemical microenvironment. Furthermore, the HOT technique allows for the creation of any desired geometry, so that user-defined, complex structures can be realized and force probed. Even dynamic rearrangement of the trapping pattern to apply forces is possible. Besides magnesium ions, other actin binding proteins or in general other substances like microtubules, intermediate filaments or DNA, can be added via the multi-channel microfluidic chamber.

The combination of high-speed video microscopy with holographic multi trap systems allowed a more accurate calibration of the optical traps than with standard methods and will in future also allow for viscoelastic measurements in a high dynamic range.

The setup is also suited to perform ensemble measurements of the whole network, like active or passive rheological investigations [228, 229]. Hereby, it would be desirable to incorporate free, untrapped particles into the network, which could act as rheological probes. An example of such a network without tracer particles is shown in figure 9.8. When having an additional, free particle in the center of this structure, it will be possible to measure rheological responses of a purely two-dimensional structure for the first time, by applying the same techniques already used in the high-speed calibration of the optical traps. Here, it would be interesting to investigate the change in the viscoelastic properties of such a network when exposed to cross-linkers, to molecular motors and to shear stress during the experiment.

A drawback of the current system is, that it is very difficult to define exact numbers of involved molecules so that single molecule measurements are challenging to realize. To get more insight into molecular interactions between single actin molecules and bundling reagents, simpler geometries have to be chosen. In the following chapters, methods to measure forces that are responsible for zipping of two filaments will be presented. This can be seen as the basic step of bundle formation in polymer systems.

Conclusions on Two-Dimensional Actin Networks

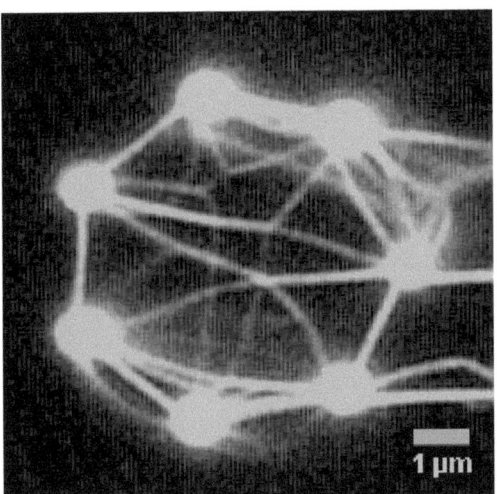

Figure 9.8: A ring like actin network structure in HOT, cross-linked by 100 mM Ca^{2+} ions. Incorporating tracer particles into the structure, could allow to investigate the viscoelastic properties of the structure using microrheological techniques.

Chapter

10

Zipping Forces Between Filaments in HOT

Parallel bundles of actin filaments are the basis of stress fiber formation, but also in actin networks bundled structures appear and some cross-linking molecules are known to promote bundle formation as well as orthogonal cross-linking, depending on the cross-linker concentration [33, 144, 150, 152]. Forces that attract filaments to each other to form bundles are similar to simple cross-linking processes. For actin binding proteins specific interactions play a major role, but also charge shielding effects according to polyelectrolyte theories can lead to condensation of filaments [227, 230–232]. To complement the experiments of the previous chapter, we wanted thus investigate the attractive forces that occur between two single filaments in the presence of bundling reagents.

In literature, there exist different methods used to investigate bundling forces between filaments, using optical tweezer techniques and theoretic approaches [37, 38, 233, 234]. In this work, the forces in a freely suspended system between trapped particles will be investigated. Based on the system presented in the last chapter (9.1), thus a protocol was created, where two filaments are arranged between three trapped beads in the microfluidic environment; see figure 10.1 for a schematic representation of different working steps. When bundling reagents are present and enough free filament is accessible, zipping occurs between the filaments. Since the fibers are confined by the trapped beads, a typical Y-shape of the zipped part of the filaments, attached to one bead and the free ends attached to the other two beads will form. Moving the traps to provide more free filament length supports the zipping process. To measure the process of free zipping, the outer traps are released subsequently. The filaments zip and the beads are pulled towards the zipping point. This process will be termed subsequently as a snapping process since the beads spontaneously snap together after being released. They will feel a frictional force of the viscous drag during the motion according to Stokes law: $F = 6\pi\eta r v$. By tracking the motion of the beads it is thus possible to define the force acting on them, since frictional force and pulling force are in equilibrium, when neglecting inertia forces, as can be done for microscopic objects. The force produced by the zipping process can thus be quantified without the need for calibrating the traps.

10.1 Experimental Procedure

The setup used for this experiment is similar to the previous experiment, presented in chapter 9.1. The identical flow cell, containing seven channels has been used (see figure 9.2). The filling of the channels was also comparable. Channel 1 contained the cross-linking solution, $MgCl_2$ at a concentration of 100 mM, in D-Buffer. Channel 2, filled with water, was employed as a drain to remove used solutions out of the flow cell during the experiment. Channel 3 was the designated microreactor, again it was blocked with an air bubble to stabilize the system. Channel 4 and channel 5 contained actin filaments, diluted in D-buffer to a final concentration of 10 nM. The concentration of actin was chosen lower than in the experiments in the previous chapter to guarantee the adsorption of single filaments to the beads. Having two channels containing actin allowed multiple consecutive experiments also, even bleaching of the actin in one of the channels. Channel 6 contained the bead suspension (1 μm, polylysine coated) at a concentration of 0.03%. In this type of experiment, there was no need for a channel containing only buffer, since the trapped beads are transferred directly into cross-linking solution.

The experimental procedure was as follows: The microreactor was filled with cross-linking solution from channel 1, leaving an air bubble between the buffer and the water in channel 3. Beads were added from channel 6 after retraction of the buffer into the drain (channel 2). The beads were trapped in a trigonal pattern with a spacing of about 10 μm and relocated to channel 3 by moving the microscopes stage. The bead suspension was removed as well, using channel 2 as drain. Subsequently, the actin solution was added from channel 4 and the beads were moved out of the microreactor into the solution. Two actin filaments were collected using fluorescence microscopy and stretched between the beads using viscous forces to align the fibers;

Experimental Procedure

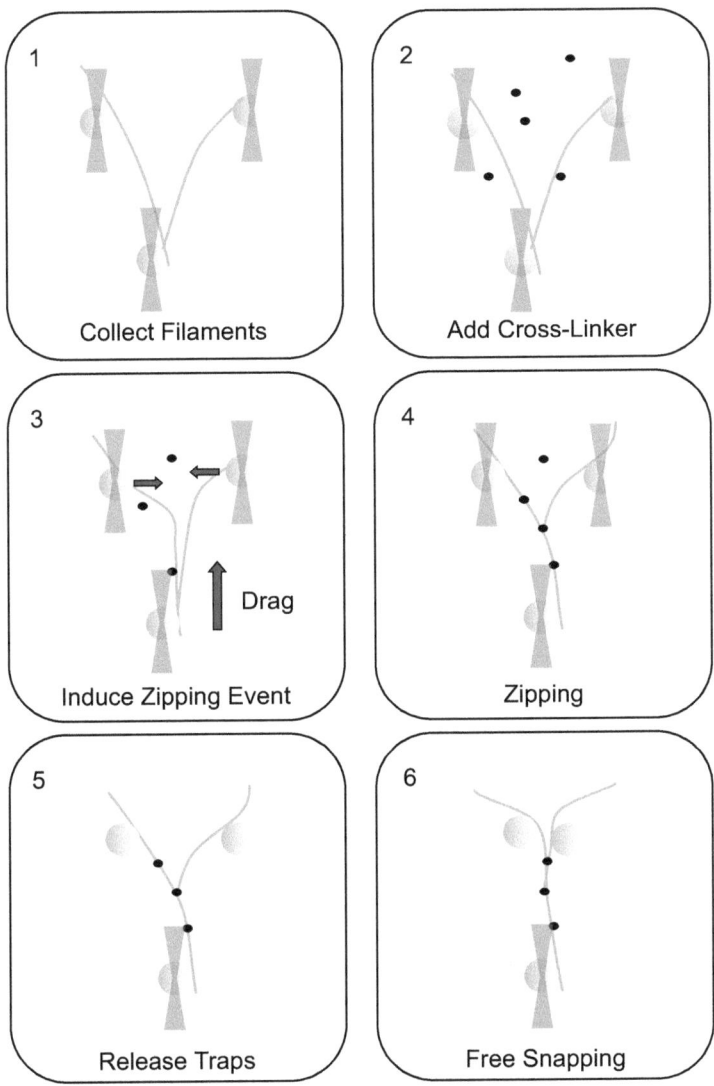

Figure 10.1: Schematic drawing of the different working steps to create a zipping between the two filaments. Three beads are held in the optical traps as is depicted by the red triangles. All steps are done inside the microreactor of the flow cell after collecting the filaments in the actin solution. Drag is produced by moving the microscopes micrometer stage. Approaching the beads to induce zipping and releasing of the two upper traps is done by updating the trapping hologram.

see figure 10.1, image 1. This step has to be done quickly to minimize photo damage on the actin. Furthermore, one has to take care to collect only two filaments, since more filaments would influence the zipping process due to bundling and and a resulting stiffening of the partners. The beads together with the attached filaments were returned to channel 3 into the cross-linking solution. To provide free filament, the outer beads of the triangle were moved then inward, by updating the trapping hologram using the SLM; see figure 10.1, images 3-4. The filaments between the beads were now free to fluctuate and bend. Using the micrometer stage, a flow was applied parallel to the filaments axis to approach the actin fibers. As soon as they touched, zipping occurred until the equilibrium of the Y-shape between the beads was reached. Since this step usually occurs during motion of the stage, no force measurement is performed at that time. Following this step, the highspeed camera was started at a frame rate of 3 kHz and the hologram was updated to release the outer traps; see figure 10.1, images 5-6. The frame rate of the fluorescence camera was set to 20 fps to record the dynamics of the filaments of the zipping.

Normally, the zipping process took less than one second. Longer time periods usually indicated the presence of multiple filaments between the beads. In this case, the zipping process was hindered doe to a higher bending stiffness of the multi-fiber bundles. Those events were not analyzed. Furthermore only events where the zipping occurred in the xy-plane were kept for analysis. Figure 10.2 gives an example of such a zipping event recorded by fluorescence microscopy. It is possible to see, how the released beads are pulled towards the zipping point, as filaments snap together. Figure 10.3 shows the corresponding position data for the beads as obtained by tracking the three particles after releasing the traps.

10.2 Data Analysis and Results

Relating the motion or the trajectories of the particles to driving force in not as straightforward as one would assume. The position data set has the form

$$r(t) = [x(t), y(t)] \tag{10.1}$$

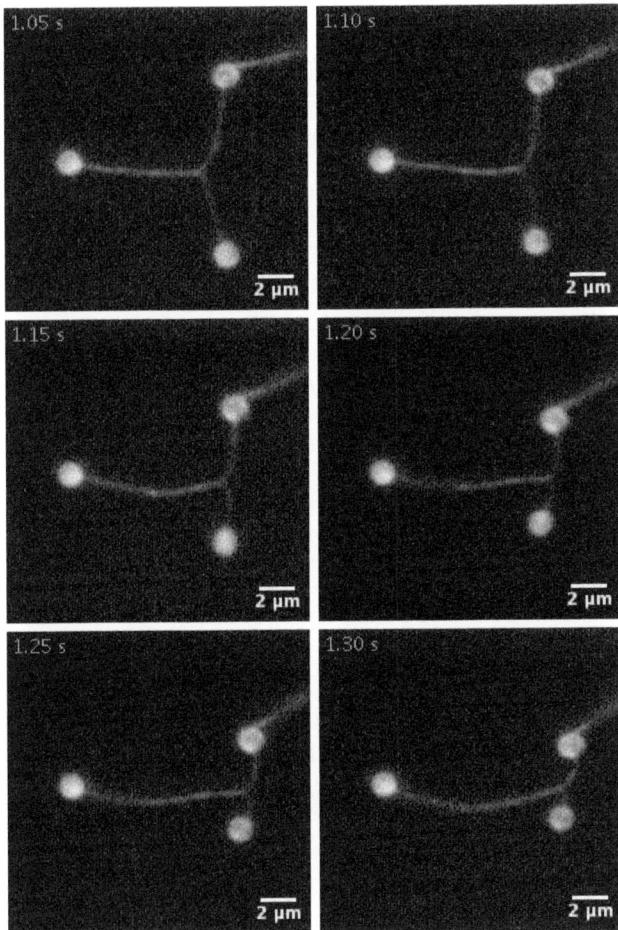

Figure 10.2: Fluorescence micrographs of the actin zipper closing spontaneously after releasing the two beads on the right. In the first image, the traps are still active, the zipper is in an equilibrium defined by the stiffness of the filaments. Removing the traps by updating the hologram causes a spontaneous very quick zipping process. The whole snapping process takes less than 300 ms.

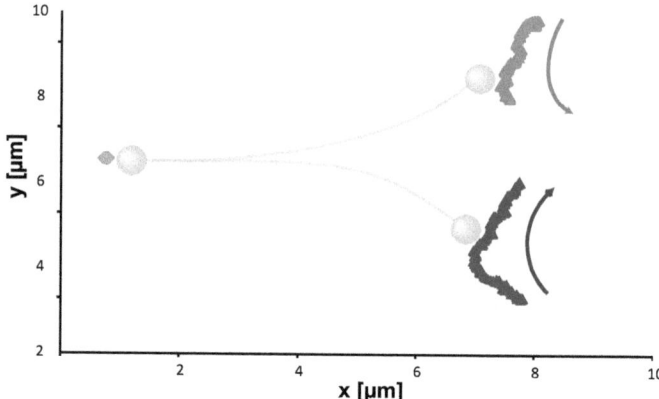

Figure 10.3: The figure shows position data of the tracked particles during snapping of the filaments. The left particle is kept in position by the optical trap. The arrows indicate the direction of motion after release of the two outer traps. The data is acquired with a frame rate of 3000 Hz over a time span of 400 ms.

i.e. the coordinates of the particle in the xy-plane at each time t. Nevertheless, velocity data can not be obtained from the simple form:

$$v_1 = (r_2 - r_1)/\Delta T. \tag{10.2}$$

Although this would result in a velocity data set, the obtained values would be dependent on the time ΔT between the measurements. The reason is, that the particles undergo brownian motion in addition to the directed movement due to the driving force. Therefore, the data has to be analyzed stochastically to yield diffusion coefficients and drift rates [228, 235]. The mean square displacement (MSD) of the particle trajectories is taken as a quantitative characteristic of the motion. It can be written as:

$$\rho(t) = \left\langle [r(t + \Delta t) - r(t)]^2 \right\rangle. \tag{10.3}$$

Data Analysis and Results

From the experimental data recorded at time intervals ΔT, the MSD can be expressed in terms of discrete time sequences:

$$\rho_x(n\Delta T) = \sum_{i=0}^{N} \frac{(x_{i+n} - x_i)^2}{(N+1)} \tag{10.4a}$$

$$\rho_y(n\Delta T) = \sum_{i=0}^{N} \frac{(y_{i+n} - y_i)^2}{(N+1)} \tag{10.4b}$$

and

$$\rho_n = \rho(n\Delta T) = \rho_x(n\Delta T) + \rho_y(n\Delta T), \tag{10.5}$$

where N defines the maximum length of the time interval. It should be chosen not to long in order to be not influenced by low frequency noise of the system. For a freely diffusing particle, the coordinates r(x,y) can be represented by a Gaussian process with transition probability P:

$$P(r|r',t) = \frac{1}{4\pi Dt} \exp\left[-\frac{(r-r')^2}{4Dt}\right] \tag{10.6}$$

and the MSD is a linear function of t:

$$\rho(t) = 4Dt. \tag{10.7}$$

When a flow or drift with a constant velocity v is superimposed, the transition probability between two coordinates becomes:

$$P(r|r',t) = \frac{1}{4\pi Dt} \exp\left[-\frac{(r-r'-vt)^2}{4Dt}\right] \tag{10.8}$$

and accordingly:

$$\rho(t) = 4Dt + v^2 t^2. \tag{10.9}$$

By fitting the experimental MSD to equation 10.9 it is possible to determine v and D. The MSD as a function of time for one exemplary particle in the snapping experiment is depicted in figure 10.4. From the velocity v we can get now the driving force F using Stokes law: $F = \gamma v$. The friction of the filaments will be neglected so that we can use the friction coefficient $\gamma = 6\pi\eta r$ of the microparticles. This force

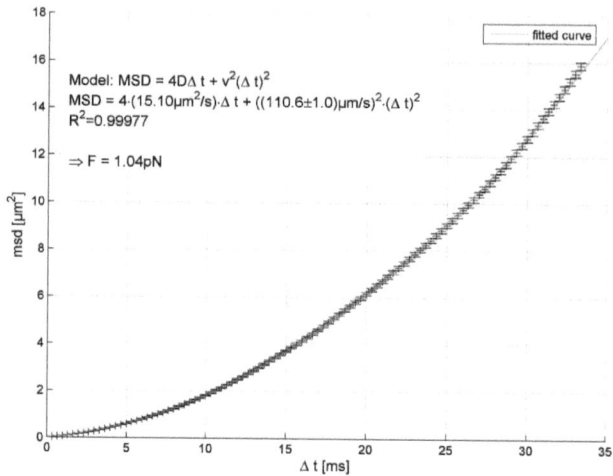

Figure 10.4: The figure shows the mean square displacement of the particle trajectory during the snapping process, after release of the trap. The red line is the fit, according to equation 10.9. Using the parameters of the particle, $r = 0.5\,\mu m$, a driving force of $1\,pN$ can be calculated.

can be understood as an adhesive energy per filament length W [234]. The force measured for both particles has to be summed up to get the total energy. This way, an adhesive energy of $2 \cdot 10^{-18}\,J/\mu m$ can be calculated. This corresponds to $1.3\,k_B T$ per actin monomer.

10.3 Conclusions on Actin Snapping Experiments

The system presented in this chapter is an application of holographic optical traps to measure intermolecular forces without the need of calibrating the single traps. This provides an extra degree of flexibility in the use of holographic optical tweezers since spring constants of single traps could vary significantly between holograms which renders force measurements with HOT often cumbersome.
A tool to measure direct molecular adhesion energies during bundling processes of

actin filaments using magnesium as a cross-linker has been developed in this part of the dissertation. The setup provided the possibility to probe single filaments, as could be proven by fluorescence microscopy and characteristic zipping behavior. Using the microfluidic environment, we could also guarantee that the zipping only happened between the desired filaments at the desired time. By the use of high-speed video microscopy, we were able to follow the complex dynamics of the particle trajectories during the zipping process. This data may also be used to get further insights in the molecular interactions of the zipping and the influence of filament bending on the process. Corrections of the adhesion energy due to filament stiffness can be expected when the bending curvature $1/R_{co}$ is in the range of $(W/l_p k_B T)^{1/2}$ [236]. Here, R_{co} is the contact curvature radius at the zipping point, it is the inverse of the radius of a circle that could be fitted to the zipping section as can be seen in figure 10.5. l_p is the persistence length of the filament and $l_p k_B T = \kappa$ is the bending rigidity of the filament. For single filaments this corresponds to a radius R_{co} of 140 nm which could not be resolved by fluorescence microscopy. Nevertheless, for stiffer filament bundles this effect is observable and explains the findings.

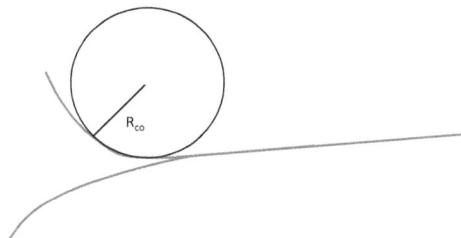

Figure 10.5: Contact curvature radius for two filaments at the zipping point.

Forces obtained in the snapping experiments were always in the range of 1 pN. A mean value of 0.85 ± 0.25 pN was measured. When summing the forces at the two beads up we get a value of 1.7 pN which corresponds well to the other results in this work. The binding energy of $1.3\,k_B T$ is comparable to literature values for polyelectrolyte counterion condensation [227, 237–239]. Nguyen *et al* for example found for DNA molecules a binding energy of $0.07\,k_B T$ per nucleotide. Considering, that every actin monomer has 11 negative charges compared to only one for DNA, giving a binding energy of $0.12\,k_B T$ per negative charge, this value seems reasonable.

The process of counterion condensation for actin bundling has bee proposed for a number of cross-linking molecules like calponin and MARCKS peptide [240, 241] which underlines the biological relevance of such measurements.

The experiments should also be conducted for other bundling agents to compare the results and get more insight into the molecular base of this event. The snapping experiment may also be conducted for other filaments like DNA or mircotubules since the system is very robust and easily adjustable to different components.

Chapter

11

Unzipping Forces Between Filaments in HOT

Complementary to the experiments presented in the previous chapter, we wanted to develop a method to measure unzipping forces between two filaments quantitatively. Unzipping is the inverse reaction of the zipping discussed in the last chapter and so, energy or a force must be provided to reverse this process. In a first attempt, the multi-channel system was also employed to perform this experiments. Holographic optical tweezers were used to produce zipper structures inside the microfluidic system as shown in chapter 10. But instead of releasing the traps, the holograms were updated to produce a pulling force on the zipper, in order to open it and determine the required forces. The protocol developed to unzip filaments between optically trapped particles is schematically depicted in figure 11.1.

11.1 Background for the Induced Unzipping of Semiflexible Polymers

The unzipping process between two fibers is determined by the elastic properties of the filaments and the involved adhesion strengths. The bundling process is always a competition between the attraction potential and the configurational entropy of the filaments (see also section 3.1.6). For semiflexible polymers, as the investigated actin filaments, the bending energy of the polymer cannot be neglected [35, 242]. Stiff polymers with high persistence lengths show a reduced loss in entropy due to

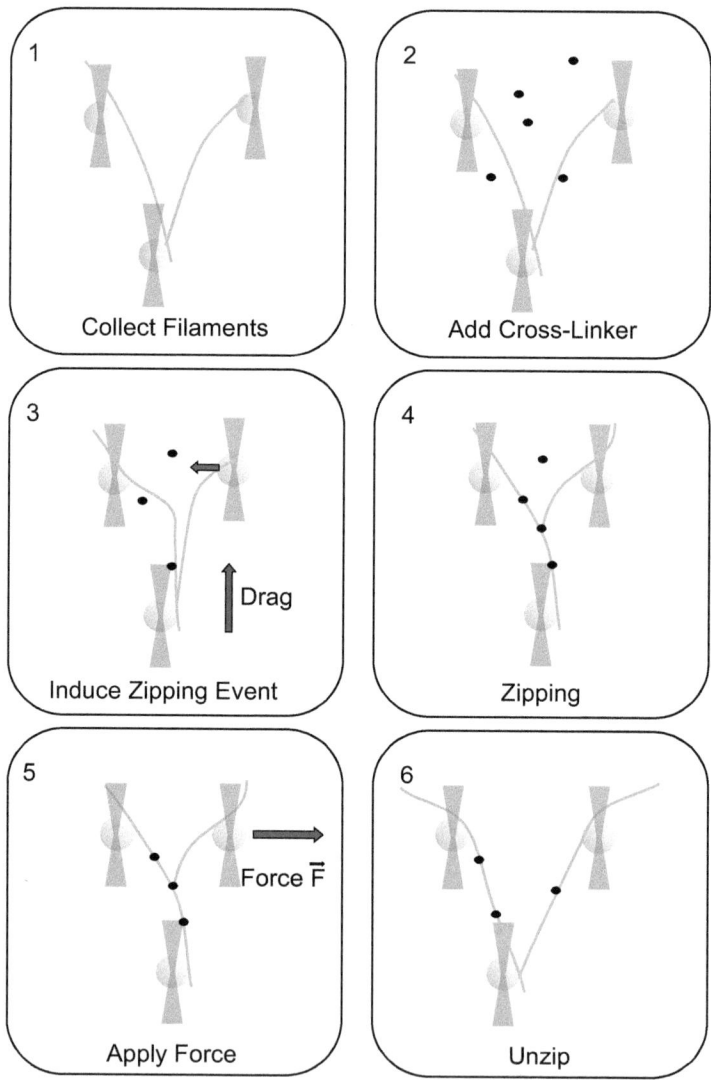

Figure 11.1: The figure shows the different working steps to unzip filaments using HOT. All steps are done inside the microreactor after collecting the filaments in the actin solution. Drag is produced by moving the microscopes micrometer stage. Approaching the beads to induce zipping and applying a force to unzip the filaments is done by updating the hologram.

bundling resulting in a preference for the bundled state. The same holds true, when the thermal motion of the polymers is restricted e.g. due to surface adsorption. Kierfeld could show theoretically, that the force induced desorption of filaments is assisted by thermal fluctuations of the polymers [234]. Thus, free fluctuating polymers will show significantly different results in unzipping experiments as compared to surface bound filaments [38, 243, 244]. For semiflexible polymers, the force-induced unzipping requires thermal activation over a characteristic energy barrier ΔG_b that scales with the square root of the bending rigidity κ.

$$\Delta G_b = \frac{2^{3/2}\kappa^{1/2}|W|^{3/2}}{3F_d}. \tag{11.1}$$

W describes here the adhesive energy per length of adsorbed filament and F_d is the desorption force. The energy barrier is a generic bending rigidity effect and gives rise to an enhanced stability against external forces. For desorption of the polymers at a constant desorption force F_d, the energy barrier ΔG_b has to be overcome by thermal activation, resulting in an Arrhenius-type desorption rate:

$$k_d = \frac{1}{\tau}e^{-\Delta G_b/T} \sim \frac{1}{\tau}e^{-F_0/F_d}, \tag{11.2}$$

where τ is a characteristic time scale of the polymer dynamics.

$$F_0 = l_p^{1/2}|f_W|^{3/2}T^{1/2} \tag{11.3}$$

is a characteristic force for the system, depending on the persistence length l_p, the free energy of adsorption f_W and the temperature T. The energy barrier gives rise to activated unzipping kinetics and leads to an enhanced dynamic stability of the bundled form for semiflexible polymers compared to flexible polymers, where no energy barrier is present. This enhanced stability should play an important role in the bundling kinetics for cytoskeletal networks. Therefore, suitable setups for the investigation of force induced unbundling transitions are of great interest in biophysical studies.

11.2 Experimental Procedure

The setup used for this experiment is similar to that used in the previous chapters (9.1 and 10.1). The same flow cell, containing seven channels was used (see figure 9.2). Filling of the channels was also identical. Channel 1 contained the cross-linking solution in D-buffer, $MgCl_2$ at a concentration of 100 mM. Channel 2, filled with water, was employed as a drain to remove used solutions out of the flow cell during the experiment. Channel 3 was the designated microreactor, again it was blocked with an air bubble to stabilize the system. Channel 4 and channel 5 contained actin filaments at a concentration of 10 nM. Channel 6 contained the bead suspension (1 μm, polylysine coated) at a concentration of 0.03%.

The experimental procedure was as well comparable to the snapping experiments in the previous chapter: The microreactor was filled with cross-linking solution, leaving an air bubble between the buffer and the water. Beads were added after retraction of the buffer into the drain. The beads were trapped in a trigonal pattern with a spacing of 10 μm and relocated to channel 3 by moving the stage. The bead suspension was removed into the drain. Subsequently, actin was added and the beads were moved out of the microreactor into the solution in the pressure compensation channel. Two actin filaments were collected using fluorescence microscopy and stretched between the beads using viscous forces to align the fibers. Again, care has to be taken to collect only two filaments. More filaments on the beads could influence the unzipping process and lead to the measurement of artifacts. Beads together with attached filaments were returned to the microreactor, into the cross-linking solution. To provide free filament, one of the beads was moved inward by updating the hologram displayed on the SLM (see figure 11.1, image 3). Moving the micrometer stage, a flow was applied parallel to the filaments axis to approach the two actin fibers. As soon as they touched, zipping occurred until the equilibrium of the Y-shape between the beads was reached. Following this step, the same bead that was used to approach the filaments was moved stepwise to apply a force on the zipping point.

Figure 11.2 shows an example of such an unzipping event. The upper of the three particles was moved by updating the hologram in 10 nm steps. 1000 steps were calculated before the experiment and updated using labview routines. To calculate the forces acting on the beads, the positions for each hologram had to be determined without force and the difference to the positions in the experiment were measured. Figure 11.3 shows the data for the tracked positions of the particles during the

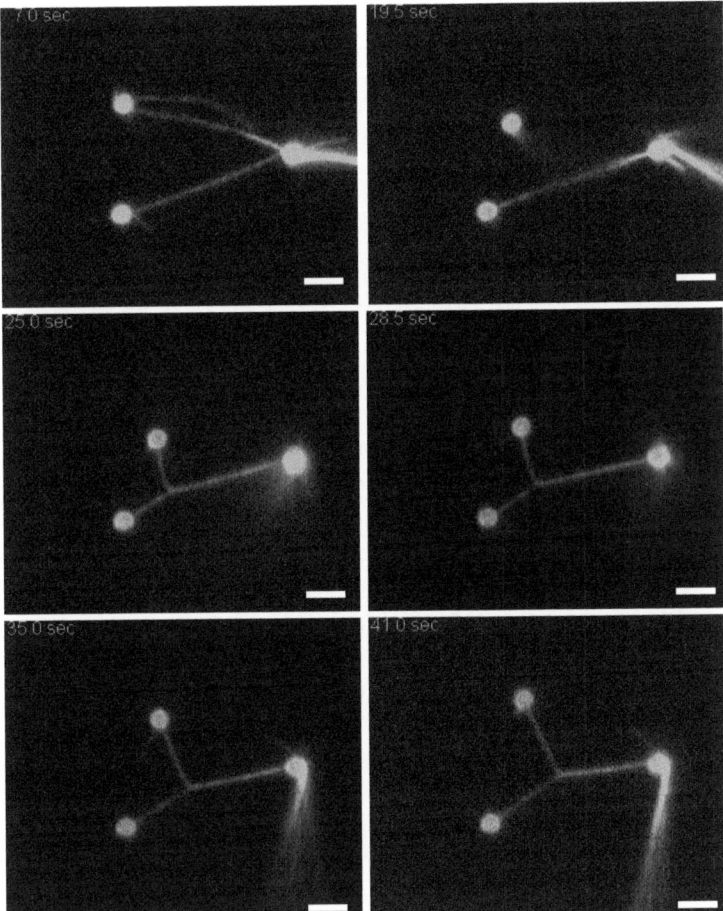

Figure 11.2: Fluorescence image time series of an experiment to unzip actin filaments bundled by magnesium ions. The upper bead was moved in 10 nm steps until the filaments zipped. Subsequently, it was retracted again, unzipping the bundle. Scale bar is 2 μm.

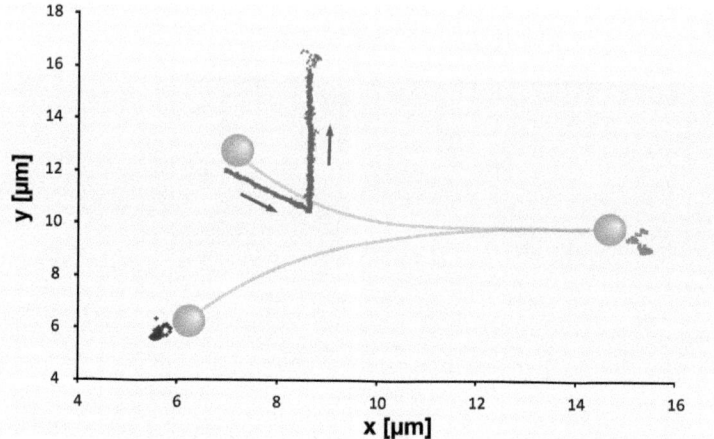

Figure 11.3: Position data set of the trapped particles in figure 11.2. The upper particle was approached to induce zipping and subsequently retracted to open the zipper again. The red arrows depict the direction of the motion.

experiment, acquired at a frame rate of 100 Hz.

11.3 Data Analysis and Results

The step size of the bead motion is crucial for this experiment but also poses the biggest challenge for the data analysis. Choosing too big steps results in a very abrupt force application. Such high loading rates could lead to unzipping forces that overcome the trapping strength of the tweezers. Furthermore, the bead is pulled out of the trap during change of the holograms, since the SLM does not provide any laser intensity in the traps during the time of the update. Smaller step sizes result in difficulties to resolve the steps in the particles trajectories. Step finding algorithms need to be applied to detect the single steps in the beads movements induced by the hologram. The response of the unzipping events is superposed to this motion, which also leads to steplike trajectories. Automated updating of the hologram at fixed update rates is a way to facilitate data analysis. However, additional challenges are posed by the requirement to calibrate all relevant holograms, since deviations in spring constants of the optical traps are present even between very similar trapping patterns [245].

A way to reduce these problems is to track the displacement of the two parti-

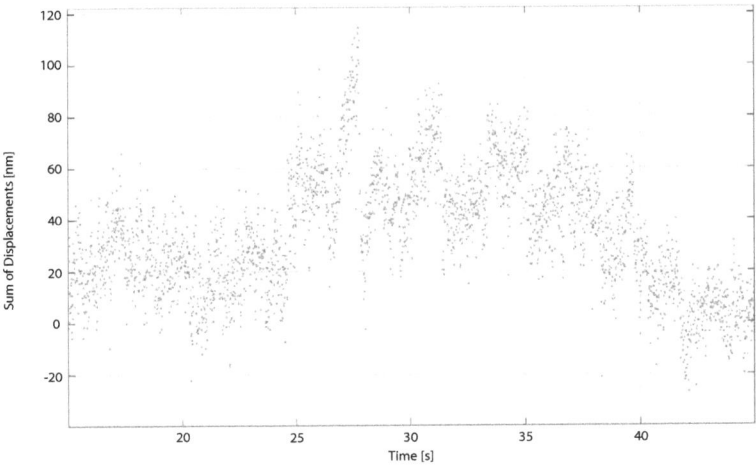

Figure 11.4: The graph shows displacement of the two beads which were not moved by the hologram during the unzipping event. For a better resolution and signal to noise ratio, the displacements of the single particles have been summed up for each frame.

cles not moved by the holograms during unzipping of the filaments. Their spring constants should not vary to much between single hologram steps. Moreover, their equilibrium position is supposed to stay constant. Such a displacement data set is shown in figure 11.4. The data show the displacement for the particles projected onto the filament axes and summed up for both particles. Therefore, only interaction transduced over the filaments was measured enhancing the signal to noise ratio and leading to a better force resolution. Forces in the order of 0.5 - 0.8 pN were measured for both beads together. When considering a similar force on the third bead, then forces for the unzipping of magnesium bundled actin filaments range between 1 and 1.5 pN. However, in most experiments the unzipping force exceeded the maximum trapping strengths of the optical tweezer and beads were pulled out of the trap.

11.4 Conclusions on Actin Unzipping Experiments using HOT

A method to unzip freely suspended actin filaments on optically trapped beads in a microfluidic environment has been developed. Complementary to the experiments in the previous chapter, cross-links between adherent filaments were actively opened up by an applied force, a key experiment when investigating the properties of such a connection. Gaining insight into such interactions will help to understand the biophysical background of bundling and unbundling events *in vivo* and how cells can actively tune these processes. Nevertheless, due to the mentioned restrictions of the holographic optical tweezers this method exhibits some drawbacks in the exact measurement of unzipping forces. Even though the two tracked particles were not moved by the update of the holograms, slight shifts in the positions were recorded between different holograms even when no filaments were attached. This could be explained by interference patterns of higher order that are known to produce unwanted laser intensity in the focal plane. Such ghost traps are known to change the intensity and exact positions of desired trapping patterns. These complications could negatively influence the force measurements. Moreover, the variance of the spring constants for all traps between different holograms poses some challenges for the data analysis. This makes it necessary to calibrate all used holograms and to assign all holograms to the exact position data in the time series.

To overcome these restrictions, optimization of the HOT setup to reduce the influence of unwanted diffraction maxima of higher order would be desirable. Spatial filtering of the laser beam, improved hologram calculation algorithms and improved, iterative SLM control to overcome these limitations have been proposed in literature [246, 247].

To avoid these problems without the need for a major remodeling of the setup, an alternative method was developed. Here, a single optical tweezer without the need for any holographic beam shaping was employed to probe freely suspended actin zipper structures. Therefore, the optical trap was combined with a system developed by Wouter Roos and Simon Schulz to create 2D actin networks on PDMS micropillar structures. In the following chapter we will discuss the experiments to unzip actin bundles suspended between pillars. Comparable to zipper structures on holographic optical tweezers a minimum influence of surface artifacts on the cross-link can be expected while avoiding some of the drawbacks of the HOT system.

Chapter

12

Unzipping Forces Between Filaments on Pillar Substrates

The unzipping experiment presented in the previous chapter (11) was transferred to pillar substrate supported filament systems to allow for better reproducibility and a more accurate readout of applied forces. Actin filament networks on PDMS pillar substrates were established as a biomimetic model of two-dimensional filament structures to investigate biomechanical properties of the network in the absence of surface effects. PDMS micropillars as scaffold for actin networks have been developed by Wouter Roos and Simon Schulz [196, 248, 249]. The pillars have a height of 15 µm and a center to center distance between 5 and 11 µm. Filaments selectively attach to the functionalized pillar tops. Filaments that bridge the gap between two pillars are therefore considered as freely suspended. The filaments can be crosslinked on the pillar tops and form extended bundle networks.

In the experiments presented in this chapter, is was our goal to find zipper structures between pillars and manipulate them with optically trapped microspheres. By applying forces perpendicular to the bundles, the zippers were opened and the required forces determined. This part of the work has been done in close collaboration with Simon Schulz who works in the department Spatz on actin structures in pillar fields. The pillar system was combined with the optical tweezer fluorescence setup to investigate the mechanics of filament zippers on pillar substrates quantitatively. A pillar substrate consists of up to 100.000 single pillars, each of which provides the possibility for a zipping event. Thus, it was possible to use one pillar field to perform a multitude of measurements minimizing the risk of photo damage to the examined filaments. Figure 12.1 shows a model of an actin network attached to a pillar substrate as used in this work. A single optical trap was used to apply the force on the zipped filaments by moving the piezo stage in 2 nm steps while the trap was held constant. This way artifacts due to the step size could be minimized and the force measurement were conducted by tracking the position of the sphere and comparison to the position before application of force. The spring constant of the trap remained constant during this movement, which further facilitated the measurement.

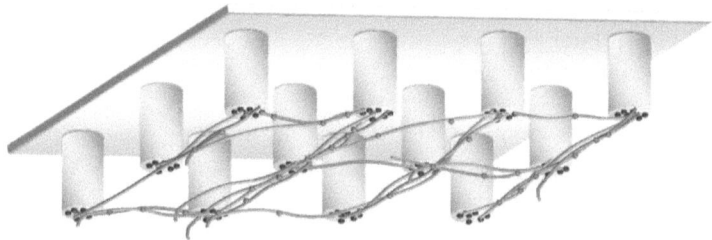

Figure 12.1: Schematic model of an actin network on a pillar substrate. The pillars were mounted upside down in the flow cell. This allowed the application of a high NA objective for trapping and imaging through the bottom glass coverslip. The pillar tops are functionalized by HMM (blue dots) to provide binding sites for actin. Cross-linker (green dots) can be added after formation of the network to bundle the filaments. The filaments are freely suspended on the pillar tops. Picture by courtesy of Timo Maier.

12.1 Experimental Procedure

The experiments on micro pillar substrates were performed in single channel flow cells as described in the materials and methods part (see figure 6.2, right side). These flow cells had one inlet and one outlet and the medium had to pass the pillar field which was hanging from the ceiling of the flow cell into the medium. The bottom of the flow cell consisted of a cover slip that allowed high resolution imaging and optical trapping inside the pillar field. Solutions were pulled into the flow cell using a syringe pump to provide a steady flow. Typical flow velocities

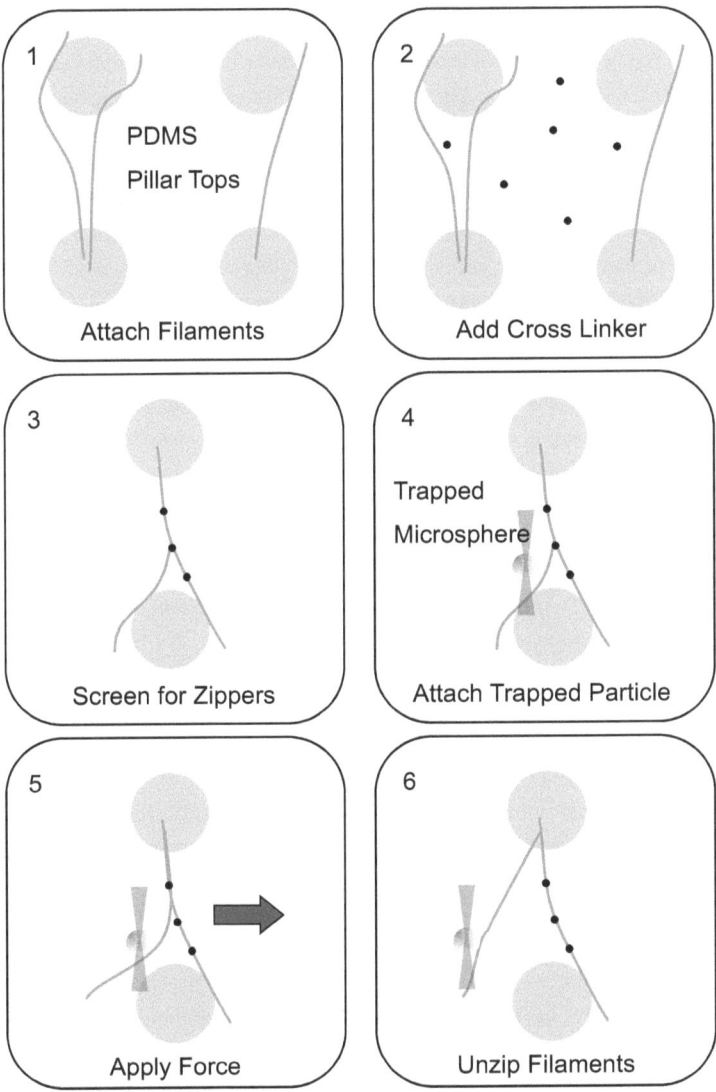

Figure 12.2: Working steps to unzip filaments on micropillar substrates: Filaments are flown into the pillar flow cell and attach to functionalized pillar tops. Cross-linker molecules are added that bundle the filaments. Zippers can form on places where the filaments are restricted by their attachment points. Beads are added and attached to zipper structures using the optical tweezers. A force is applied by moving the stage relatively to the trapped particle. The zipper will open up and the required force can be determined.

were in the order of 0.5-2 µl/min to allow adhesion of molecules on the pillar tops. The pillar tops were functionalized by HMM as described in Materials and Methods (see section 6.6 for details). After rinsing with D-buffer, the actin filaments were flown in at a concentration of 25 nM in D-buffer. The flow cell was rinsed again with D-buffer, once a sufficient number of filaments were attached to the pillar tops. Subsequently cross-linker was added, diluted in D-buffer. Magnesium ions as cross-linker were added at concentrations of 20 mM and α-actinin at 750 nM. By fluorescence microscopy the formation of the bundling could be observed. Thereupon the poly-(L-lysine) functionalized microspheres (2 µm diameter, polystyrene) were flown into the cell at a dilution of 0.02 %. The flow was stopped to trap a sphere in a single optical trap. For this purpose, the spatial light modulator was used as a simple reflecting surface not modifying the wave front of the laser. The pillar field was screened for zipper structures using fluorescence microscopy. Once a zipper was located, the trapped bead was attached onto the filament close to the zipping point. Subsequently, the microscope stage was moved at 0.2 µm/s to apply an unzipping force on the zipper. The position of the attached particle was recorded using the brightfield microscope at a frame rate of 500 frames per second during the motion. This way the unzipping force could be directly measured, since the attached filament exerted a force on the bead in the stationary trap. A scheme of the different steps during this experiment is shown in figure 12.2. The trap was calibrated using Brownian motion calibration and power spectra analysis after each experiment.

A fluorescence image series of such a force induced unzipping event is presented in Figure 12.3. A zipper like structure was attached to two pillars indicated by dashed circles on the right and the top of picture 1. An optically trapped bead was attached to the free end of the filament. The stage, including the pillar substrate was moved to the right while the trap remained at a fixed position. The binding sites between the filament were successively released and the zipper opened up. By tracking the position of the trapped particle and comparing it to the equilibrium position of the trap it was possible to obtain force curves for the unzipping process. Every time a bond is broken, a drop in the measured force could be observed due to released filament between two adjacent binding sites. A force curve belonging to the experiment of figure 12.3 is shown in figure 12.4. The steps in the force curve can be assigned to the pictures 2-5. Three steps of comparable force can be seen. The drop in the force plateau at t=19.5 s can be attributed to a release of a long piece

Experimental Procedure 135

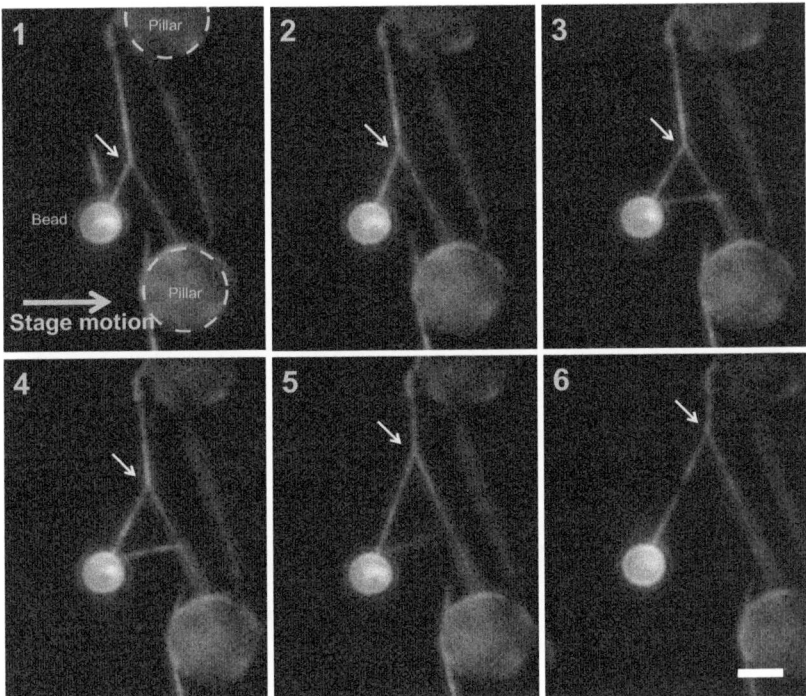

Figure 12.3: Unzipping of two filaments attached to pillar substrate. A bead is attached to a zipper structure between pillars (**1**). The stage is moved to the right, the zipper opens up and the cross-linking point propagates along the filament (**2-5**) Yellow arrows indicate the position of the zipping point. Scale bar is 2 µm.

of filament between image 4 and 5. So it is possible to show, that the cross-linkers are not equally distributed along the filament.

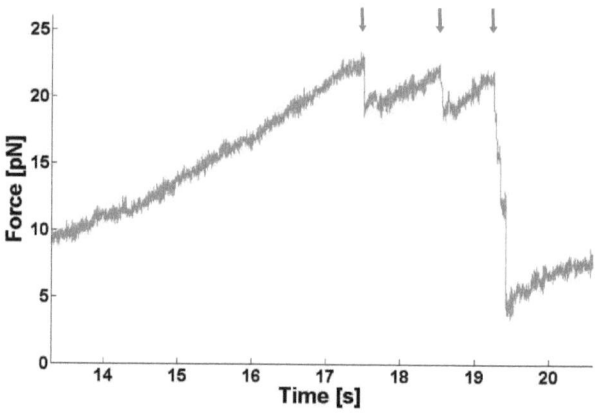

Figure 12.4: Force curve for unzipping process. Three successive steps are visible as indicated by red arrows. It is possible to see the increase in force before each drop. The slope of the curve was used to obtain the loading rate of the respective measurement. Steps correspond to figure 12.3.

12.2 Data Analysis and Results

Compared to other experiments measuring cross-linking forces on actin filaments, our experiment allowed the simultaneous fluorescence imaging of the whole unzipping process [37, 38]. This is especially helpful to distinguish artifacts like detachment of the filament from the particle from real unzipping events, since this would be very hard to judge just by the force curves. Also effects like unfolding of the cross-linker, that could lead to a drop in the force curve can unambiguously be excluded, when correlating the force curves with the fluorescence imaging. An example of such a detachment of the filament from the bead is shown in figure 12.5. Using this additional information, allows to exclude such events from the data analysis easily. Another aspect to be considered is the directionality of the applied force. The force can only be transmitted via the filament, that is attached to the trapped particle. This force will always have a component parallel to the cross-link and a perpendicular one. Only the perpendicular forces act on the cross-link, therefore the measured forces have to be corrected by the sine of the angle between the two filaments. This correction is only possible with the presented setup, where both filaments are in the focal plane of the imaging device.

When measuring molecular rupture forces one will always obtain a broad distribu-

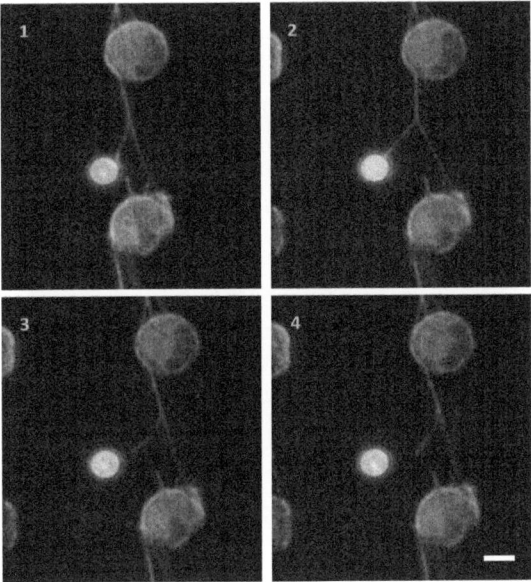

Figure 12.5: Detachment of filament from trapped bead. Such an event would give a similar drop in the force curve as an unzipping event. So the fluorescence imaging allows to distinguish such events. Scale bar is 2 μm.

tion of possible values and the most frequent force will depend how fast the bond is loaded. This is due to a thermal activation of the bonds that lead to lower forces for low loading rates and vice versa [250, 251]. To compare the results of different measurements it is necessary to determine the loading rate of the experiment. Thus, the slope of the increase in pulling force just before an unzipping event was determined by a linear fit (see inset of figure 12.6). From the slope of the force curve the loading rate can directly be determined. In principle, the loading rate is a function of the retraction velocity of the microscope stage, but since the investigated filament system consists of elastic elements a control had to be performed. Figure 12.6 shows a histogram of measured loading rates. They are distributed between 2 and 8 pN/s and therefore all in the same order of magnitude. The rupture force depends on the logarithm of the loading rate and a significant influence on the rupture forces is only observed if this value ranges over several orders of magnitude [252, 253]. Therefore, it is possible to compare the different measurements.

Figure 12.7 and figure 12.8 show a comparison of the measured forces for actin cross-

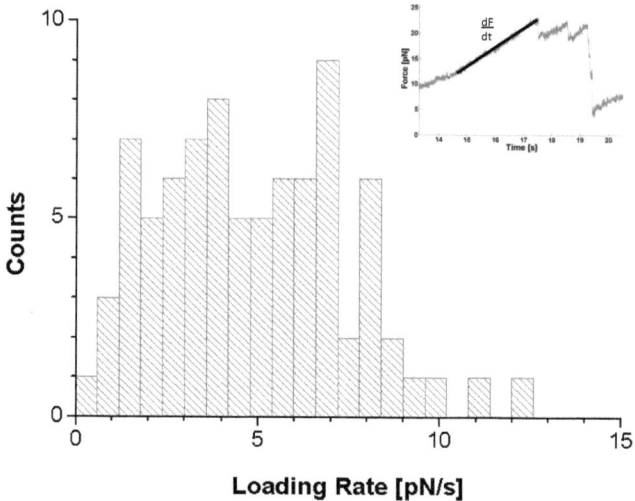

Figure 12.6: Histogram of measured loading rates during unzipping events of actin filaments bundled by 20 mM magnesium ions. Forces are equally distributed between 2 and 8 pN/s. The inset shows how the loading rate was determined from the recorded force curves, n=82.

linked by 20 mM Mg^{2+} and by 750 nM α-actinin. For magnesium as cross-linker all measured forces were below 30 pN, while in the case of α-actinin typical unzipping forces between 30 and 45 pN were observed. This increase can be explained by the higher specificity of the actin-actinin interaction that leads to higher rupture forces. The values are comparable to those known from literature [37, 38]. However, the risk of measuring surface artifacts is definitely reduced in our setup. To our knowledge this is the first system where it is possible to correlate the measured forces directly to the propagation of the zipping point along the filament. The broad distribution of measured unzipping forces might be attributed to thermal motion of the filaments, an effect that could not be investigated in other setups yet.

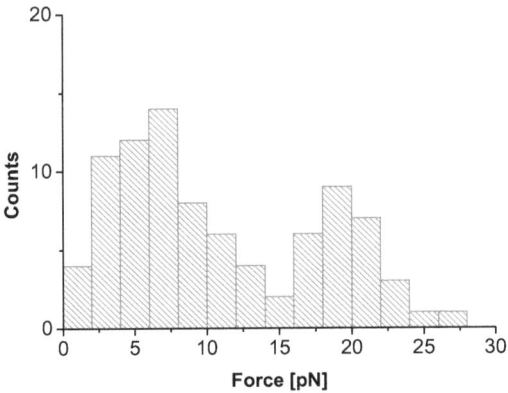

Figure 12.7: Histogram of measured forces during unzipping of actin cross-linked by 20 mM Mg^{2+}, n=82.

12.3 Conclusions on Actin Unzipping Experiments on Pillar Substrates

We could show that it is possible to combine pillar substrates with the optical tweezers technology. The unzipping experiments presented in the last chapter could be significantly extended. Constructing the zipper arrays on the pillar substrates allowed the investigation of a multitude of filaments in one flow cell. The applicable forces of the tweezers could be increased due to the prevention of holographically modified optical tweezers without reducing the possible complexity of the investigated system. Also the data readout was facilitated by this measure. Surface artifacts could be excluded in the freely suspended zipper structures. Also artifacts due to detachment of the filaments from pillars or the bead can be excluded in the data evaluation when taking the fluorescence images as control. Moreover, the effects

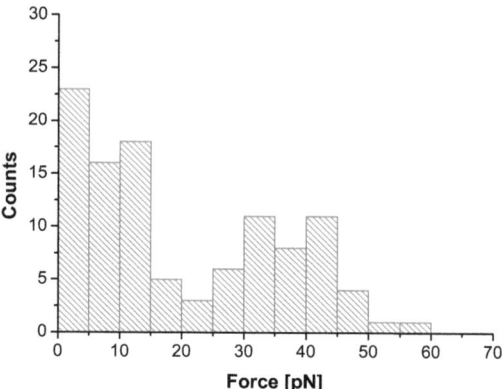

Figure 12.8: Histogram of measured forces during unzipping of actin cross-linked by 750 nM α-actinin, n=101.

of thermal motion of the filaments on the unzipping forces should be much better observable in our setup than in comparable ones. Such effects might be involved also in the unzipping process in biological systems and are topic of ongoing research [254]. As a first example, we investigated the unzipping forces of actin filaments cross-linked by 20 mM Mg^{2+} solution and by 750 nM α-actinin. A significant difference in the required rupture forces was observable. However an exact analysis of the different force histograms would be desirable to get more insights into the nature of unzipping events between semiflexible polymers. This could include an analysis of the cooperative effects of adjacent binding sites, since one would expect that not only the last bond is affected by a pulling force on the filament. Therefore, more experiments at different concentrations of cross-linkers might be helpful to reduce the decoration of the filaments with adhesive molecules. This way it would also be possible to screen for potential rebinding effects during the unzipping process. Also experiments at different loading rates could give further insights into the process of the filament unzipping process.

Chapter

13

Probing Adhesion of *Plasmodium* Sporozoites with Optical Tweezers

Gliding motility of apicomplexan parasites depends on formation of adhesions to the surface that are necessary to mediate forces to an intracellular motility machinery based on actomyosin interactions. In *Plasmodium* sporozoites, a group of transmembrane proteins is responsible for the connection to the surface [255]. TRAP (Thrombosponin Related Anonymous Protein) is the most prominent member of this group, thus it is named the TRAP protein family. Parasites lacking certain proteins from this group either fail to invade salivary glands of the mosquito, have a decreased infectivity due to reduced motility inside the mammalian skin or show difficulties in invading the liver of the host [40, 173, 177, 256, 257]. The different members of the TRAP family consist of similar adhesive domains [190], so their function is partially redundant and they can substitute each other to a certain degree. However, the exact role of most of its members during the complex process of adhesion, motility and invasion is still largely unknown. Also the function of actin in this process remains ambiguous. Actin is not only required for motility but also the adhesion shows a dependency on filamentous actin. Adhesion of sporozoites depends not only on the presence of adhesive proteins on the surface of the sporozoite, also an active process to bring the parasite into contact with the surface seems to be important. Using optical traps we wanted to probe the different steps of the adhesion-locomotion process of *Plasmodium* sporozoites in order to examine more closely the different partners involved. Using optical tweezers, adhesions are induced by pressing sporozoites actively onto the glass surface. The strength of the formed adhesion in turn is probed by applying pulling forces. Finally, it is possible to investigate motile behavior of the parasites by manipulation with optical traps.

13.1 Viability of Sporozoites in Optical Traps

In order to investigate living objects with optical traps, the survival of the specimen must be guaranteed. Optical tweezers are often designed to use wave lengths in near IR, since this spectral window is known to show very little absorption in aqueous samples from biological objects [54, 258, 259]. An easy method to reduce photo damage is the use of beads as handles, that could be trapped and attached to the object, so that the laser has not to be focused into the specimen [260]. However, since we wanted to investigate adhesive properties of the sporozoites, the use of adhering particles was discarded to reduce the risk of observing artifacts due to adhesion signaling by the particle. Thus, a method to monitor the viability of the sporozoite during the experiment was developed. SYTOX Orange is a fluorescence dye whose fluorescence intensity increases 450 fold upon intercalation with DNA [261, 262]. But since cell membranes are not permeable to the dye, staining will only occur in dying or dead cells where the membrane is at least partially degraded. The absorption maximum of SYTOX Orange is at 547 nm and the emission maximum at 570 nm, so the green fluorescence laser already implemented in the setup could be used for imaging. Clear differences between healthy ad dead sporozoites in the fluorescence images allowed the determination of sporozoite viability in the laser traps (see figure 13.1). The concentration of the fluorescence dye was chosen such, that also in healthy sporozoites the nucleus was visible by a very faint fluorescence signal. This allowed the determination of the apical end of the sporozoite, since it is known, that the nucleus is always located in the backmost third of the sporozoite. It could be shown, that 87% of sporozoites survived more than 200 s inside the laser trap, therefore only experiments, that were finished within this time span were taken for subsequent analysis (see figure 13.2).

13.2 Force Calibration for Trapped Sporozoites

Calibrating the spring constant for a trap holding an asymmetric sporozoite is not as straight forward as for spherical objects like beads. Since the sporozoite is much

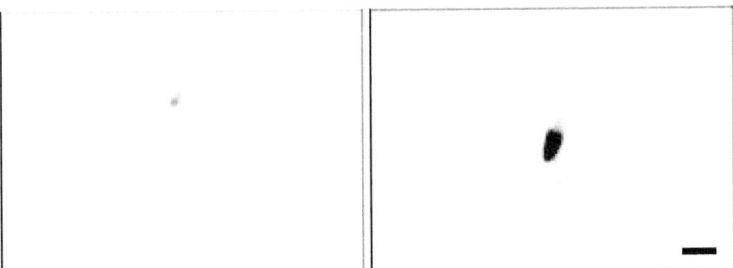

Figure 13.1: Viability test for *Plasmodium* sporozoites using SYTOX Orange. Inverted fluorescence micrographs are shown for better contrast. On the left side a healthy sporozoite is shown; only a very faint fluorescence signal from the nucleus is observable. On the right side the same sporozoite when dead: The fluorescence signal from the nucleus is much stronger. Also a weak fluorescence signal in the cell plasma due to fragmented DNA is visible. Scale bar is 2 µm.

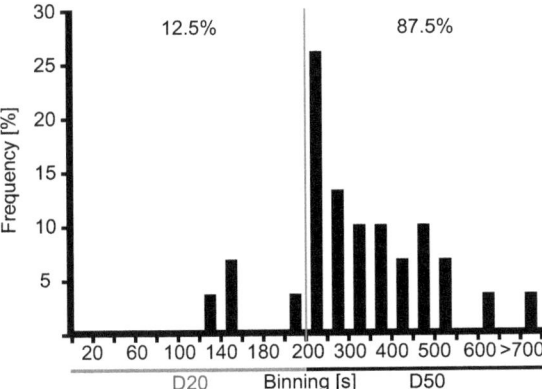

Figure 13.2: Viability histogram for *Plasmodium* sporozoites in laser trap at 100 mW. Relative frequencies for mortality events are given. 200 s are chosen as maximum experiment time to guarantee survival of sporozoites (red line in diagram). Note the change in the binning size at $t = 200$ s for better readability. Sample size $n = 31$.

bigger than the focus of the laser, the laser can be moved in the volume of the object along the axis for some micrometers without causing significant forces. Thus, simple calculations based on brownian motion are not leading to the correct results. Nevertheless using a calibration methodology based on hydrodynamic friction forces allows to find the maximum trapping forces at which the objects escapes the trap [263].

The microscope stage was moved at increasing velocities until the hydrodynamic friction of the medium exceeded the trapping strength and the object was pushed out of trap. To calculate the associated forces, the friction coefficient of the object must be determined. For spherical objects, friction can be calculated by the equation:

$$F_{Sphere} = 6\pi\eta r v. \tag{13.1}$$

η is the viscosity of the medium, r is the radius of the sphere and v is the relative velocity. Sporozoite are elongated, curved bodies. Therefore, corrections according to Perrin friction factors (Jean Baptiste Perrin, 1870-1942) could be conducted [264–266]. These correction factors are multiplicative adjustments to the friction of deformed spheroids relative to the corresponding friction of perfect spheres of the same volume. The Perrin friction factor was calculated for a prolate spheroid, i.e. a spheroid with one long and two short axes **a** and **b** (see figure 13.3). The axial ratio p is thus:

$$p = \frac{a}{b}. \tag{13.2}$$

For better readability the Perrin S factor is defined as:

$$S = 2\frac{\operatorname{arctanh}\xi}{\xi}, \tag{13.3}$$

having the parameter ξ:

$$\xi = \frac{\sqrt{|p^2 - 1|}}{p}. \tag{13.4}$$

So the Perrin correction factor for translational friction can be derived as:

$$f_P = \frac{2p^{2/3}}{S}. \tag{13.5}$$

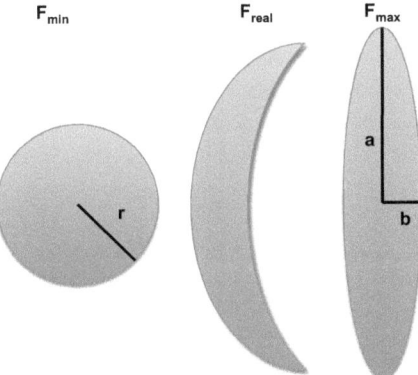

Figure 13.3: Model for Perrin friction factor calculation for *Plasmodium* sporozoites. The Sporozoites are considered as prolate spheroids, having one long ($a = 5\,\mu m$) and two short axes ($b = 1\,\mu m$). Radius of a sphere with identical volume as a sporozoite is calculated as $r = 1,34\,\mu m$, since the average volume of a sporozoite is $10\,\mu m^3$ as derived from cryo electron tomography data (Misha Kudryashev, *unpublished results*). The real shape and thus the real friction force acting on the sporozoite F_{real} can be considered in between these two extrema.

Accordingly, the force acting on a spheroid is given as:

$$F_{Spheroid} = f_P \cdot F_{Sphere}. \tag{13.6}$$

Using equations 13.1 and 13.6 it is thus possible to calculate the maximum friction force, at which the trapping force is exceeded. This is also the maximum force, one can apply on the object with the optical tweezer. Nevertheless, one has to bear in mind, that the sporozoite is not a stiff object. When being exposed to friction forces, it will bend and adjust its orientation to minimize the hydrodynamic friction. So the calculated force gives only an upper limit of the potential trapping force F_{max}, whereas the force calculated for an isovoluminous sphere would give a lower limit for the friction force on the sporozoite F_{min}. The real friction force acting on the sporozoite can be considered to be between these two extrema. Escape velocities were measured for different laser powers and the respective forces for sphere and spheroid were calculated. For 100 mW trapping power, the minimal force F_{Sphere}

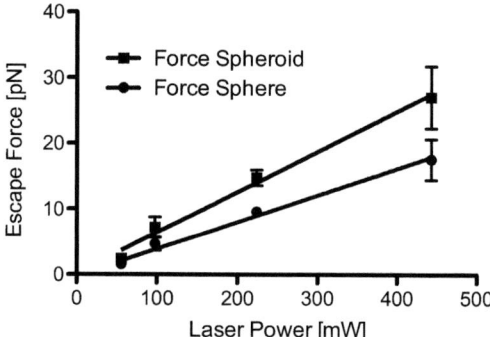

Figure 13.4: Escape forces for trapped sporozoites at different laser powers. For each laser power 7 sporozoites were probed. Stage velocity was increased linearly until the sporozoite was pushed out of the trap. Error bars are the standard deviation of measurements. The lower values indicate the escape force as calculated for a sphere and the higher ones the forces calculated for a prolate spheroid of sporozoite proportions.

was determined by linear regression as $1,5 \pm 0,2$ pN and the upper limit $F_{Spheroid}$ as $2,4 \pm 0,4$ pN. For the maximum applied laser power of 450 mW the respective values were $F_{Sphere} = 17 \pm 3,0$ pN and $F_{Spheroid} = 27 \pm 4,7$ pN. Figure 13.4 shows a comparison of escape forces calculated by this model for different laser powers.

13.3 Adhesion Experiments with *Plasmodium* Sporozoites

The establishment of a first adhesion site of sporozoites is independent on the presence of actin as well as TRAP. In all samples, sporozoites underwent easily a primary adhesion, which could also not be released by the maximum force of the optical tweezers. Therefore, we focused on the second adhesion formed by sporozoites during their attachment process. Wild type sporozoites that were not under the influence of actin depolymerizing drugs showed in all cases primary and secondary adhesions, leading eventually to motility. Sporozoites lacking TRAP or wild type sporozoites lacking filamentous actin due to depolymerization by cytochalasin D showed only one adhesion site, indicating that actin may be responsible for the formation of a second adhesion. But it could be shown in flow experiments that these sporozoites would nevertheless form a second adhesion when subjected to a force, that pushes the parasite onto the substrate. Consequently the tweezers experiment was designed as following (The different steps of an experiment are shown schematically in figure 13.5): Sporozoites lacking either TRAP or filamentous actin were allowed to attach onto a glass surface. Once sporozoites had formed a single adhesion site they were examined for adhesion at the front end or back end using the fluorescence of the nucleus due to SYTOX Orange. Subsequently, they were trapped with the optical tweezers at the free end of the sporozoite and pressed against the glass surface. Using the optical tweezers it was tried to establish a second adhesion site. Hereby, all possible sides of the sporozoite were brought into contact with the surface to screen for potential adhesive loci. If no secondary adhesion was formed within 200 s - the maximum experiment time - then the sporozoite was counted as negative result for secondary adhesion. Once an adhesion was formed, it was probed by pulling the trap upwards to check for the stability of the bond (see figure 13.6 for example experiment). These experiments were performed for wild type sporozoites under the influence of cytochalasin D, for TRAP knockout parasites and for TRAP knockout parasites using cytochalasin D.

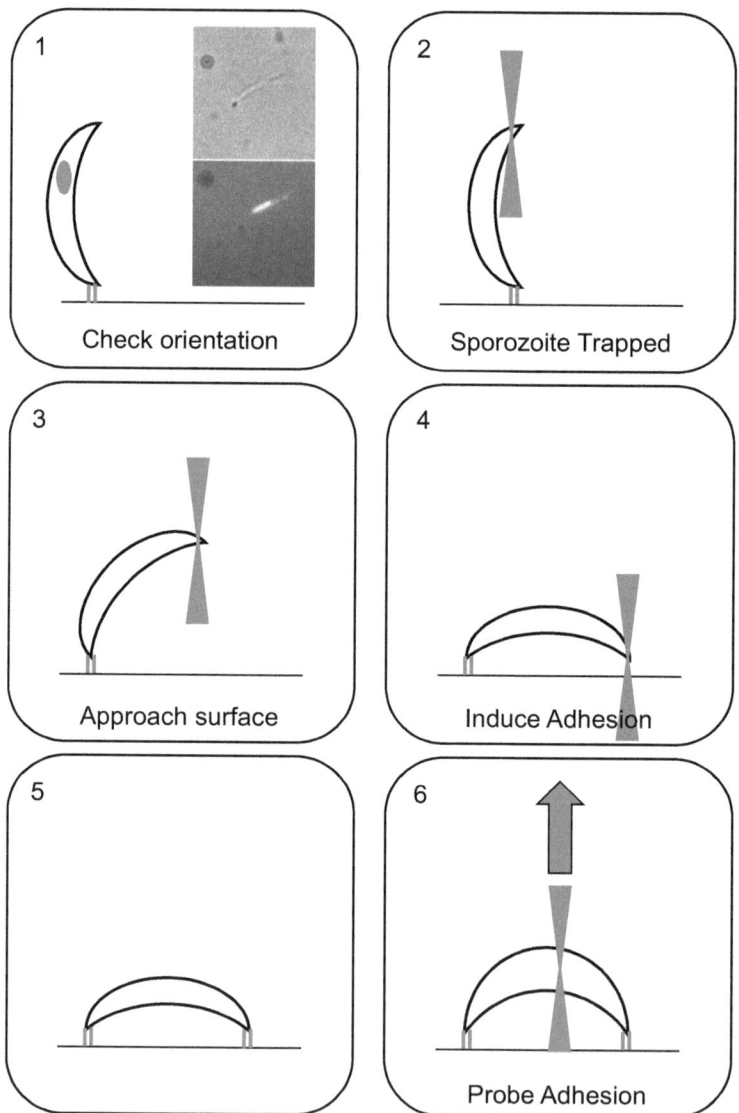

Figure 13.5: The figure shows the principle for sporozoite adhesion experiments. Only sporozoites that have already a primary adhesion are probed. The position of the nucleus and therefore the orientation of the sporozoite is determined using the fluorescence of SYTOX Orange (inset of image **1**). Using the optical trap a secondary adhesion is established and probed. Only adhesions that cannot be detached by the optical tweezers are counted.

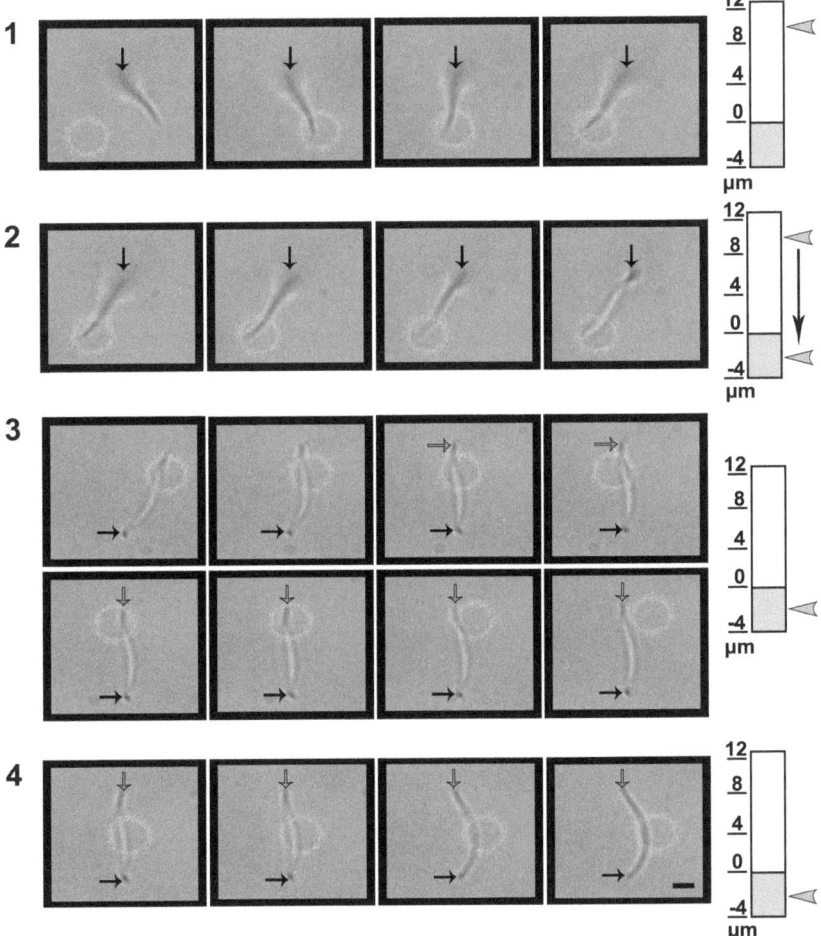

Figure 13.6: Image series of sporozoite adhesion experiment. The position of the laser trap is marked by an orange circle in the images. The black arrow indicates the primary adhesion of the sporozoite. On the right side is a scale for the focus height above the glass surface. The trap is always in focus. **1**: A rear end adhering sporozoite is trapped and moved in x and y direction. The adhesion is out of focus, about 9 microns below the focal plane (see scale on the right). **2**: The end of the sporozoite is brought into contact with the glass surface by moving the focus down. Orange arrowheads on the scale indicate motion in z-direction. **3**: The end of the sporozoite is moved in x and y on the glass surface. The trap pulls the sporozoite onto the glass. Once a second adhesion is formed (red arrow), the sporozoite stops following the trap's motion. **4**: The center of the sporozoite is moved in x-, y- and z-direction to probe the adhesion. Scale bar is 2 μm.

13.4 Results and Discussion of Sporozoite Experiments

Wild type sporozoites without actin depolymerizing drugs showed secondary adhesion in all cases. Sporozoites with only one adhesion site could not be observed in any experiments. Using cytochalasin D hardly any sporozoites formed two adhesion sites. Instead most of them were attached at their front end, having their rear end freely dangling into the medium. This is especially interesting, since motile sporozoites, that are exposed to actin depolymerizing drugs could remain attached at both the rear and front end, as revealed by additional flow experiments in the group of Friedrich Frischknecht. This suggests that primary adhesion of sporozoites is independent of actin only at the front end of the parasite. The same behavior could be observed for TRAP knockout sporozoites. Using the optical tweezers, a secondary attachment was induced in 42% of F-actin deficient parasites and 45% of TRAP knockout parasites (see figure 13.7). In the case of TRAP knockout parasites under the absence of F-actin hardly any formation of secondary adhesion could be observed, even under the active force of the optical trap. Only in 8% of parasites a secondary adhesion could be induced. To analyze the results statistically, a Fishers exact test was performed. The Fishers exact test is an specialization of the chi-square test for smaller sample sizes. It showed that the results are statistically highly significant. This indicates, that the formation of a secondary adhesion could be dependent on an active force that might be generated by actin or TRAP involving signaling pathways. However it seems, that TRAP and F-actin can substitute each other in the presence of an external force, so that secondary adhesion could still be observed.

13.5 Conclusions and Outlook for *Plasmodium* Experiments

We developed a method to probe mechanical influences on the adhesion process of *Plasmodium* sporozoites. The survival of the parasites during the experiments could be guaranteed and a range for applicable forces was defined. It could be shown, that in the absence of either filamentous actin or TRAP, the force of the optical tweezers

Conclusions and Outlook for *Plasmodium* Experiments

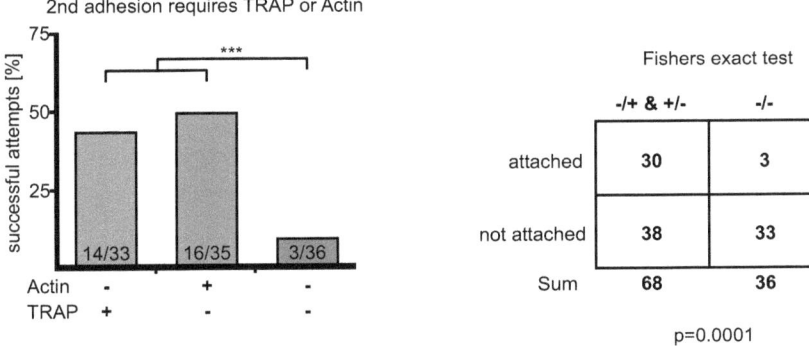

Figure 13.7: Formation of secondary adhesion induced by optical tweezer. On the left side diagram for induced secondary adhesion under the influence of cytochalasin D and and TRAP knockout. Absolute numbers of experiments are written in the diagram. On the right side results of the fishers exact test. Wildtype without F-actin and TRAP knockout with normal F-actin are summed up as one population against TRAP knockout without F-actin.

could induce the formation of secondary adhesions. In the absence of an external force no secondary adhesion was formed; a result that was confirmed by additional flow experiments in the group of Friedrich Frischknecht. This is a clear hint, that a chemomechanical coupling is involved in the formation of a secondary adhesion and that a concerted interaction of actin and TRAP is involved in this process. In the absence of actin and TRAP no secondary adhesion at all could be established, even under the influence of the force of the optical trap. This proofs, that at least one of this partners must be present. Thus we were able to show a molecular link for the adhesion process. Neither for wild type sporozoites nor for TRAP knockout and F-actin deficient parasites, it was possible to pull the parasite off the substrate, indicating that adhesive forces exceed 30 pN in this cases. However, more genetically modified parasites could be probed to investigate the properties of different adhesion proteins. Therefore, further experiments on this system would be worthwile.

Additionally, the applications of optical traps can be used to investigate the motility of sporozoites. Patch gliding sporozoites, isolated from the hemolymph of mosquitoes for example could be stalled by optical tweezers (see figure 13.8 1a and b). As well in this system force measurements and the investigation of molecular connections would be desirable. Finally, in experiments not presented here, it was possible to simulate the patch gliding behavior and pull adhering sporozoites over their adhesion site without detaching the parasite from the surface (see figure 13.8 2a and b). This could give further insights into the fascinating machinery of a gliding motility adhesion.

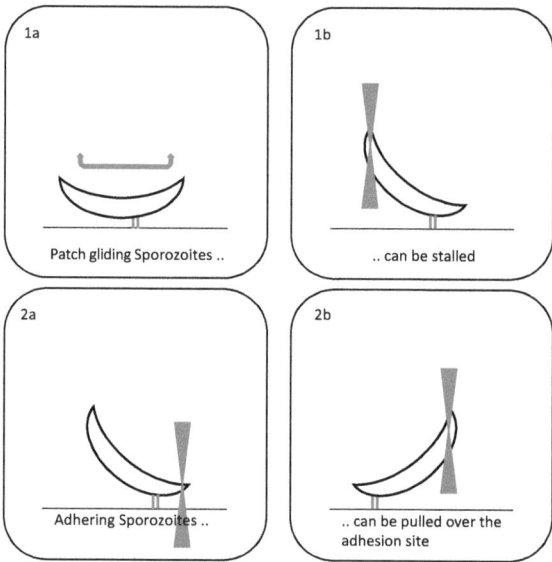

Figure 13.8: Further experiments, that are possible with optical tweezers. Probing the mechanism and the properties of patch gliding sporozoites could give new insights into the gliding motility machinery. Probing the properties of a slipping adhesion could help to understand the connection between the adhesion site and the intraparasitic cytoskeleton.

Chapter

14

Discussion and Outlook

This work aimed to develop systems that allow for the qualitative and quantitative investigation of biomimetic cytoskeletal structures. A holographic optical tweezers device was combined with a multi camera microscopy setup, to create, image and mechanically probe biomimetic actin structures. High-speed video microscopy was used to calibrate multi-trap systems, using power spectra analysis of particle motion. This allows for reliable calibration of the optical traps even in the presence of low frequency noise, especially present in microfluidic systems. A microfluidic system was developed, that provided the controlled exchange of the chemical environment of optically trapped structures without exceeding the trapping strength of the optical traps by hydrodynamic friction forces. Thus, high resolution force measurements with subpiconewton resolution during dynamic events like cross-linking of actin networks were possible. Two-dimensional actin networks between optically trapped polystyrene microspheres in the microfluidic environment could be constructed and subsequently cross-linked. This system represents a biomimetic model of the actin cortex of cells and the first purely two-dimensional protein network that has been mechanically probed. We were able to measure the contractile forces during cross-linking of actin by magnesium ions, which to our knowledge is the first quantitative force measurement experiment with holographic optical tweezers. We found that the contraction of an actin network between seven optically trapped spheres in a hexagonal pattern with $14\,\mu$m diameter exceeded energies of 600 $k_B T$, giving an insight into the forces that cells could produce by cross-linking their cortical cytoskeleton. The presented system would be as well suited for rheological studies on such two-dimensional protein structures or in general as a versatile force sensor array in a microfluidic environment.

Holographic optical tweezers were also used to investigate the zipping process of two actin filaments in the presence of magnesium ions as bundling agent. Zipper structures between filaments that were attached to three optically trapped microparticles were created. After release of the traps, the zipper closed completely against the viscous drag of the adherent particles. By tracking the particles trajectory with high-speed video microscopy, it was possible to calculate the friction forces and the respective adhesion energy of the two filaments.

In a variation of this experiment the holographic optical tweezers were used to unzip two filaments, which opens up a way to directly measure unbundling forces of actin filaments.

Finally, the unzipping experiment was transferred to actin network structures suspended on micropillar arrays. The incorporation of optical tweezers into this system allowed for quantitative force measurements. We were able to unzip filaments attached to the pillars with an optically trapped particle and record the force response curves. This combination of the two technologies greatly improved the capacities of the individual systems. The pillar substrates could be extended to quantitative force measurements and the tweezers approach was improved with respect to reproducibility and read-out possibilities compared to the HOT approach to unzip filaments. The unzipping of two freely suspended filaments offers significant advantages over existing approaches for the measurement of rupture forces between cross-linked filaments where filaments are attached to a glass surface. In our setup, both filaments are free to fluctuate between the pillars. This setting is much closer to the *in vivo* situation, where thermal motion and bending modes influence the bundling behavior. Moreover, in our setup, the whole experiment is performed in the xy-plane, which gives a higher degree of control by fluorescence imaging of the whole structure during all steps of the unzipping process.

This work describes the development of a toolbox for the investigation of cytoskeletal multicomponent systems. The microfluidic approach allows the complete control of the chemical environment of the system without affecting the ability to perform high-resolution force measurements. Thus, extended structures like filament bundles and filament networks can become the object of chemo mechanical investigations. As a first example for the versatility of the setup we probed actin filaments cross-linked by magnesium ions and α-actinin, but the system can be easily adapted to other cross-linker systems or polymers like microtubules, intermediate filaments or DNA. Furthermore, it would be interesting to improve the system by incorporation of more advanced fluorescence imaging techniques. This could be multi-color fluorescence for the simultaneous imaging of filaments and cross-linkers or high-resolution fluorescence techniques like stimulated emission depletion microscopy (STED) to resolve multiple filaments in bundles or the curvature radius of filaments at the zipping point.

In a second experiment sporozoites of the malaria causing parasite *Plasmodium* have been probed with optical tweezers. Sporozoites use a special adhesion and gliding machinery, based on adhesive transmembrane proteins and an intracellular actomyosin system. The influence of external forces on the adhesion process were investigated. This is the first time that sporozoites were probed with optical tweezers, so a systematic protocol for trapping experiments has been established.

We probed the formation of secondary adhesion sites of the parasite. Experiments were performed for wild type sporozoites, sporozoites under the influence of actin filament disrupting drugs and knock-out parasites, lacking TRAP, a protein known to be involved in adhesion processes. We could find clear evidence for the involvement of TRAP and filamentous actin in the formation of a secondary adhesion as well as for the supportive role of an external force in adhesion formation. In our experiments this external force was mimicked by optical tweezers, but might be provided *in vivo* by hydrodynamic forces generated by flow in blood vessels.

Sporozoites show a unique motility behavior that is based on a highly adaptive protein machinery. Therefore, they are a well suited model organism to investigate mechanical influences on adhesive behavior. It could be shown that the optical tweezers are an ideal tool for mechanical single cell experiments on sporozoites. Further experiments involving different knock-out parasites and quantitative force measurements on adhesion and motility of the *Plasmodium* parasite will give valuable insights into molecular mechanisms that might lead to the development of new drugs interfering with the infection process of malaria.

List of Figures

2.1 Image of the comet Lulin approaching the earth in spring 2009 14
2.2 Photons hitting a surface A in the time interval Δt 15
2.3 Image of a Crookes radiometer, also known as light mill 17
2.4 Schematic representation of trapping setups realized by Arthur Ashkin 18
2.5 Diffraction of an incident ray by a dielectric sphere 20
2.6 Change in the momentum of an incident laser beam of Gaussian shape on a dielectric sphere. 21
2.7 Gradient forces and scattering forces on a dielectric sphere as a function of incidence angle 23
2.8 A typical power spectrum for a trapped particle 32
2.9 Fourier Transform of an incident light wave by a lens 38
2.10 A phase mask encoding a hexagonal trap pattern 39
2.11 Schematic drawing of a liquid crystal phase modulator in reflection mode 40
2.12 Setup for holographic optical tweezers using a spatial light modulator 41
2.13 Prism hologram displacing an optical trap in the focal plane to a point $\vec{p}$. 43

3.1 Schematic summary of the mechanism of actin polymerization 49
3.2 Schematic model of a myosin II molecule 50
3.3 Peptide motifs of α-actinin 51
3.4 Zipping of two actin filaments by cross-linking molecules 53

4.1 Life cycle of *Plasmodium* parasite 58

4.2	Model for sporozoite gliding machinery	60
4.3	Proteins of the TRAP family found in *Plasmodium*	62
4.4	Image sequence, motility of *Plasmodium* sporozoites	62
4.5	Schematic of multi-step adhesion of *Plasmodium* sporozoite	63
5.1	The complete microscopic setup used in the experiments	68
5.2	Fluorescence excitation laser modulation	70
5.3	Example micrographs obtained with the different imaging setups	73
5.4	Screen shot of the software platform to control the setup	74
6.1	Passivation of silicon substrates	80
6.2	Channel flow cell and pillar flow cell	82
7.1	Scanning confocal micrograph of actin filaments attached to a surface	88
8.1	Green fluorescent *Plasmodium* sporozoites	92
9.1	Model of a two-dimensional fiber network on trapped beads in a microfluidic environment	97
9.2	Example of flow scheme to add and retract different solutions in a multi channel flow cell	98
9.3	Two-dimensional actin network between trapped beads	101
9.4	Calibration results obtained by Boltzmann statistics and power spectra analysis	103
9.5	Contraction of hexagon due to cross-linking forces, brightfield images before and after cross-linking	105
9.6	Build up of contractile forces on beads during cross-linking of the hexagon	106
9.7	Total energy stored in the optical traps during cross-linking of the hexagon	107
9.8	Ring like actin network structure in HOT	109
10.1	Working steps to measure zipping forces by free snapping of filaments	113
10.2	Time series of zipping occurring after trapped beads are released	115
10.3	Position data of tracked beads during filament snapping	116
10.4	Mean square displacement of particle trajectory during snapping	118
10.5	Contact curvature radius for two filaments at the zipping point	119

List of Figures

11.1	Working steps to unzip filaments using HOT	122
11.2	Time series of filament unzipping, using HOT	125
11.3	Position data of particles during unzipping using HOT	126
11.4	Displacement of beads during unzipping	127
12.1	Scheme of an actin network on a pillar substrate	132
12.2	Working steps to unzip filaments on micropillar substrates	133
12.3	Unzipping process of filaments on micropillar substrate	135
12.4	Force curve for unzipping process	136
12.5	Detachment of filament from trapped bead	137
12.6	Histogram of measured loading rates during unzipping events	138
12.7	Histogram of measured forces during unzipping of actin cross-linked by 20 mM Mg	139
12.8	Histogram of measured forces during unzipping of actin cross-linked by 750 nM α-actinin	140
13.1	Viability test for *Plasmodium* sporozoites using SYTOX Orange	143
13.2	Viability histogram for *Plasmodium* sporozoites in laser trap at 100 mW	143
13.3	Model for Perrin friction factor calculation for *Plasmodium* sporozoites	145
13.4	Escape forces for trapped sporozoites versus laser power	146
13.5	Principle for sporozoite adhesion experiments	148
13.6	Image series of sporozoite adhesion experiment	149
13.7	Formation of secondary adhesion induced by optical tweezer	151
13.8	Outlook for sporozoite experiments with optical tweezers	152

Bibliography

[1] B. Alberts, D. Bray, J. Lewis, M. Raff, K. Roberts, and J. D. Watson. *Molecular Biology of the Cell*. Garland, **2002**. 9, 45, 46

[2] N. Wang, J. P. Butler, and D. E. Ingber. Mechanotransduction across the cell surface and through the cytoskeleton. *Science*, 260, (5111): 1124–1127, **1993**. 9, 45, 46, 47

[3] D.E. Ingber. The mechanochemical basis of cell and tissue regulation. *Mechanics and Chemistry of Biosystems*, 1: 58–68, **2004**. 9

[4] M. E. Cusick, N. Klitgord, M. Vidal, and D. E. Hill. Interactome: gateway into systems biology. *Human Molecular Genetics*, 14, (90002): 171–181, **2005**. 10

[5] M. Sarikaya, C. Tamerler, A. K. Y. Jen, K. Schulten, and F. Baneyx. Molecular biomimetics: nanotechnology through biology. *Nature Materials*, 2, (9): 577–585, **2003**. 10

[6] C. T. Lim, E. H. Zhou, A. Li, S. R. K. Vedula, and H. X. Fu. Experimental techniques for single cell and single molecule biomechanics. *Materials science & engineering C. Biomimetic and supramolecular systems*, 26, (8): 1278–1288, **2006**. 10

[7] V. Noireaux and A. Libchaber. A vesicle bioreactor as a step toward an artificial cell assembly. *Proceedings of the National Academy of Sciences*, 101, (51): 17669–17674, **2004**. 10

[8] T. Yanagida, M. Nakase, K. Nishiyama, and F. Oosawa. Direct observation of motion of single F-actin filaments in the presence of myosin. *Nature*, 307, (5946): 58–60, **1984**. 10

[9] J. T. Yang, W. M. Saxton, R. J. Stewart, E. C. Raff, and L. S. Goldstein. Evidence that the head of kinesin is sufficient for force generation and motility in vitro. *Science*, 249, (4964): 42–47, **1990**. 10

[10] K. Svoboda, C. F. Schmidt, B. J. Schnapp, and S. M. Block. Direct observation of kinesin stepping by optical trapping interferometry. *Nature*, 365, (6448): 721–727, **1993**. 10

[11] C. Veigel, M. L. Bartoo, D. C. S. White, J. C. Sparrow, and J. E. Molloy. The stiffness of rabbit skeletal actomyosin cross-bridges determined with an optical tweezers transducer. *Biophysical Journal*, 75, (3): 1424–1438, **1998**.

[12] A. E. Clemen, M. Vilfan, J. Jaud, J. Zhang, M. Barmann, and M. Rief. Force-dependent stepping kinetics of myosin-V. *Biophysical Journal*, 88, (6): 4402–10, **2005**. 10

[13] A. Ashkin. Acceleration and trapping of particles by radiation pressure. *Physical Review Letters*, 24, (4): 156–159, **1970**. 10, 18

[14] A. Ashkin, J. M. Dziedzic, J. Bjorkholm, and S. Chu. Observation of a single-beam gradient force optical trap for dielectric particles. *Optics Letters*, 11: 288–290, **1986**. 18, 23

[15] Jeffrey R Moffitt, Yann R Chemla, Steven B Smith, and Carlos Bustamante. Recent advances in optical tweezers. *Annual Review of Biochemistry*, 77: 205–228, **2008**. 10

[16] E. Helfer, S. Harlepp, L. Bourdieu, J. Robert, F. C. MacKintosh, and D. Chatenay. Microrheology of biopolymer-membrane complexes. *Physical Review Letters*, 85, (2): 457–460, **2000**. 10

[17] E. A. Abbondanzieri, W. J. Greenleaf, J. W. Shaevitz, R. Landick, and S. M. Block. Direct observation of base-pair stepping by RNA polymerase. *Nature*, 438, (7067): 460, **2005**.

[18] O. Thoumine and J. J. Meister. Dynamics of adhesive rupture between fibroblasts and fibronectin: microplate manipulations and deterministic model. *European Biophysics Journal*, 29, (6): 409–419, **2000**. 10

[19] E. R. Dufresne, G. C. Spalding, M. T. Dearing, S. A. Sheets, and D. G. Grier. Computer-generated holographic optical tweezers arrays. *Review of Scientific Instruments*, 72: 1810–1816, **2001**. 10, 35, 37

[20] D. G. Grier and Y. Roichman. Holographic optical trapping. *Applied Optics*, 45, (5): 880–887, **2006**.

[21] J. E. Curtis, B. A. Koss, and D. G. Grier. Dynamic holographic optical tweezers. *Optics Communications*, 207, (1-6): 169–175, **2002**. 10, 35

[22] J. Anderson, D. Chiu, R. Jackman, O. Cherniavskaya, J. McDonald, H. Wu, S. Whitesides, and G. Whitesides. Fabrication of Topologically Complex Three-Dimensional Microfluidic Systems in PDMS by Rapid Prototyping. *Science*, 261: 895, **1993**. 10

[23] M. A. Unger, H. P. Chou, T. Thorsen, A. Scherer, and S. R. Quake. Monolithic microfabricated valves and pumps by multilayer soft lithography. *Science*, 288, (5463): 113, **2000**.

[24] D. Erickson and D. Li. Integrated microfluidic devices. *Analytica Chimica Acta*, 507: 11–26, **2004**.

[25] George M. Whitesides. The origins and the future of microfluidics. *Nature*, 442, (7101): 368–373, **2006**. 10

[26] J. Enger, M. GoksâĹŽâĹĆr, K Ramser, P. Hagberg, and D. Hanstorp. Optical tweezers applied to a microfluidic system. *Lab on a Chip*, 4, (196-200), **2004**. 11

[27] C. H. Schmitz, J. Curtis, and J. P. Spatz. Constructing and probing biomimetic models of the actin cortex with holographic optical tweezers. *Proc of SPIE*, 5514: 446–454, **2004**.

[28] R. T. Dame, M. C. Noom, and G. J. Wuite. Bacterial chromatin organization by H-NS protein unravelled using dual DNA manipulation. *Nature*, 444, (7117): 387–90, **2006**.

[29] E. Eriksson, J. Scrimgeour, A. Graneli, K. Ramser, R. Wellander, J. Enger, D. Hanstorp, and M. Goksor. Optical manipulation and microfluidics for studies of single cell dynamics. *Journal of Optics A: Pure and Applied Optics*, 9, (8): 113, **2007**.

[30] E. Eriksson, J. Enger, B. Nordlander, N. Erjavec, K. Ramser, M. GoksâĹŽâĹĆr, S. Hohmann, T. NystrâĹŽâĹĆm, and D. Hanstorp. A microfluidic system in combination with optical tweezers for analyzing rapid and reversible cytological alterations in single cells upon environmental changes. *Lab on a Chip*, 7, (1): 71–76, **2007**. 11

[31] D. Bray, J. Heath, and D. Moss. The membrane-associated cortex of animal cells: its structure and mechanical properties. *Journal of Cell Science*, 20: 71–88, **1986**. 11, 52

[32] M. Bowick, A. Cacciuto, G. Thorleifsson, and A. Travesset. Universal negative poisson ratio of self-avoiding fixed-connectivity membranes. *Physical Review Letters*, 87, (14): 148103, **2001**. 11

[33] Y. Tseng. The bimodal role of filamin in controlling the architecture and mechanics of f-actin networks. *Journal of Biological Chemistry*, 279, (3): 1819–1826, **2003**. 11, 111

[34] G. T. Charras, C. K. Hu, M. Coughlin, and T. J. Mitchison. Reassembly of contractile actin cortex in cell blebs. *Journal of Cell Biology*, 175, (3): 477, **2006**. 11, 95

[35] J. Kierfeld and R. Lipowsky. Unbundling and desorption of semiflexible polymers. *Europhysics Letters*, 62, (2): 285–291, **2003**. 11, 52, 121

[36] Jan Kierfeld, Krzysztof Baczynski, Petra Gutjahr, and Reinhard Lipowsky. Semiflexible polymers and filaments: From variational problems to fluctuations. volume 1002, pages 151–185. AIP, **2008**. 11, 47

[37] H. Miyata, R.i Yasuda, and K. Kinosita. Strength and lifetime of the bond between actin and skeletal muscle [alpha]-actinin studied with an optical trapping technique. *Biochimica et Biophysica Acta (BBA) - General Subjects*, 1290, (1): 83–88, **1996**. 11, 52, 111, 136, 138

[38] J. M. Ferrer, H. Lee, J. Chen, B. Pelz, F. Nakamura, R. D. Kamm, and M. J. Lang. Measuring molecular rupture forces between single actin filaments and actin-binding proteins. *Proceedings of the National Academy of Sciences*, 105, (27): 9221, **2008**. 11, 52, 111, 123, 136, 138

[39] R. Amino, S. Thiberge, B. Martin, S. Celli, S. Shorte, F. Frischknecht, and R. Menard. Quantitative imaging of plasmodium transmission from mosquito to mammal. *Nature Medizine*, 12, (2): 220–224, **2006**. 11, 57, 59

[40] A. A. Sultan, V. Thathy, U. Frevert, K. J. H. Robson, A. Crisanti, V. Nussenzweig, R. S. Nussenzweig, and R. Menard. TRAP Is Necessary for Gliding Motility and Infectivity of Plasmodium Sporozoites. *Cell*, 90: 511–522, **1997**. 12, 141

[41] K. Matuschewski, A. C. Nunes, V. Nussenzweig, and R. Menard. Plasmodium sporozoite invasion into insect and mammalian cells is directed by the same dual binding system. *EMBO Journal*, 21: 1597–1606, **2002**.

[42] S. H. I. Kappe, K. Kaiser, and K. Matuschewski. The plasmodium sporozoite journey: a rite of passage. *Trends in Parasitology*, 19, (3): 135–143, **2003**. 12

[43] J. Kepler. *De cometis libelli tres, I. Astronomicvs, theoremata continens de motu cometarum.* AvgvstÃę Vindelicorvm, **1619**. 13

[44] W. Crookes. The radiometer and its lessons. *Nature*, 17, (418): 7–8, **1877**. 17

[45] J. C. Maxwell. On stresses in rarified gases arising from inequalities of temperature. *Philosophical Transactions of the royal society of London*, pages 231–256, **1879**. 17

[46] P. N. Lebedev. Untersuchungen über die Druckkräfte des Lichtes (Investigations on the pressure forces of light). *Annalen der Physik*, 6: 433âĂŽĂĐÄň458, **1901**. 16

[47] A. Ashkin and J. M. Dziedzic. Optical levitation by radiation pressure. *Applied Physics Letters*, 19, (8): 283–285, **1971**. 18

[48] A. Ashkin. Trapping of atoms by resonance radiation pressure. *Physical Review Letters*, 40, (12): 729, **1978**. 18

[49] S. Chu. Nobel lecture: The manipulation of neutral particles. *Reviews of Modern Physics*, 70, (3): 685–706, **1998**.

[50] S. Chu, J. E. Bjorkholm, A. Ashkin, and A. Cable. Experimental observation of optically trapped atoms. *Physical Review Letters*, 57, (3): 314, **1986**.

[51] S. Chu, L. Hollberg, J. E. Bjorkholm, A. Cable, and A. Ashkin. Three-dimensional viscous confinement and cooling of atoms by resonance radiation pressure. *Physical Review Letters*, 55, (1): 48, **1985**. 18

[52] W. Ketterle. Nobel lecture: When atoms behave as waves: Bose-einstein condensation and the atom laser. *Reviews of Modern Physics*, 74, (4): 1131, **2002**. 18

[53] A. Ashkin and J. M. Dziedzic. Internal cell manipulation using infrared laser traps. *Proceedings of the National Academy of Sciences*, 86, (20): 7914–7918, **1989**. 18

[54] A. Ashkin, J. M. Dziedzic, and T. Yamane. Optical trapping and manipulation of single cells using infrared laser beams. *Nature*, 330, (6150): 769–771, **1987**. 142

[55] A. Ashkin and J. M. Dziedzic. Optical trapping and manipulation of viruses and bacteria. *Science*, 235, (4795): 1517–1520, **1987**. 18, 31

[56] A. Ashkin. Forces of a single-beam gradient laser trap on a dielectric sphere in the ray optics regime. *Biophysical Journal*, 61, (2): 569–582, **1992**. 19, 20, 23

[57] G. Roosen. Optical levitation of spheres. *Canadian Journal of Physics*, 57: 1260–1279, **1979**. 19

[58] T. Wohland, A. Rosin, and E. H. K. Stelzer. Theoretical determination of the influence of the polarization on forces exerted by optical tweezers. *Optik (Stuttgart)*, 102: 181–190, **1996**. 20

[59] A. Rohrbach. Stiffness of optical traps: Quantitative agreement between experiment and electromagnetic theory. *Physical Review Letters*, 95, (16): 168102, **2005**. 20, 22

[60] A. Rohrbach and E. H. K. Stelzer. Optical trapping of dielectric particles in arbitrary fields. *Journal of the Optical Society of America A*, 18, (4): 839–853, **2001**. 22, 25

[61] Y. Harada and T. Asakura. Radiation forces on a dielectric sphere in the rayleigh scattering regime. *Optics Communications*, 124, (5-6): 529–541, **1996**. 23

[62] J. P. Gordon. Radiation forces and momenta in dielectric media. *Physical Review A*, 8, (1): 14, **1973**. 24

[63] L. W. Davis. Theory of electromagnetic beams. *Physical Review A*, 19, (3): 1177–1179, **1979**. 25

[64] K. Visscher and G. J. Brakenhoff. Theoretical study of optically induced forces on spherical particles in a single beam trap. I: Rayleight scatterers. *Optik*, 89, (4): 174–180, **1992**. 25, 34

[65] P. C. Chaumet and M. Nieto-Vesperinas. Time-averaged total force on a dipolar sphere in an electromagnetic field. *Optics Letters*, 25, (15): 1065–1067, **2000**. 25

[66] R. Gauthier. Computation of the optical trapping force using an FDTD based technique. *Optics Express*, 13, (10): 3707–3718, **2005**. 25

[67] F. Merenda, G. Boer, J. Rohner, G. DelacrâĹŽÂĺtaz, and R. P. SalathâĹŽÂĺ. Escape trajectories of single-beam optically trapped micro-particles in a transverse fluid flow. *Optics Express*, 14, (4): 1685–1699, **2006**. 26, 30

[68] K. Svoboda and S. M. Block. Biological applications of optical forces. *Annual Reviews in Biophysics and Biomolecular Structure*, 23: 247–285, **1994**. 27

[69] E. J. G. Peterman, F. Gittes, and C. F. Schmidt. Laser-induced heating in optical traps. *Biophysical Journal*, 84, (2): 1308–1316, **2003**. 29, 31

[70] R. C. Weast. Handbook of chemistry and physics 53rd ed. *Cleveland, Ohio: The Chemical Rubber Co*, **1972**. 31

[71] S. C. Kuo and M. P. Sheetz. Force of single kinesin molecules measured with optical tweezers. *Science*, 260, (5105): 232–234, Apr 1993. 31, 106

[72] M. C. Wang and G. E. Uhlenbeck. On the theory of the brownian motion II. *Reviews of Modern Physics*, 17, (2-3): 323–342, **1945**. 31

[73] S. Keen, J. Leach, G. Gibson, and M. J. Padgett. Comparison of a high-speed camera and a quadrant detector for measuring displacements in optical tweezers. *Journal of Optics A: Pure and Applied Optics*, 9: 264–266, **2007**. 33

[74] K. Visscher, S. P. Gross, and S. M. Block. Construction of multiple-beam optical traps withnanometer-resolution position sensing. *IEEE Journal of Selected Topics in Quantum Electronics*, 2, (4): 1066–1076, **1996**. 34

[75] E. Faellman and O. Axner. Design for fully steerable dual-trap optical tweezers. *Applied Optics*, 36, (10): 2107–2113, **1997**. 34

[76] K. Sasaki, K.a Masanori, M. Hiroaki, K. Noboru, and M. Hiroshi. Pattern formation and flow control of fine particles by laser-scanning micromanipulation. *Optics Letters*, 16, (19): 1463–1465, **1991**. 34

[77] C. Mio, T. Gong, A. Terray, and D. W. M. Marr. Design of a scanning laser optical trap for multiparticle manipulation. *Review of Scientific Instruments*, 71, (5): 2196–2200, **2000**. 34

[78] J. M. Tam, I. Biran, and D. R. Walt. An imaging fiber-based optical tweezer array for microparticle array assembly. *Applied Physics Letters*, 84: 4289, **2004**. 35

[79] Y. Y. Sun, L. S. Ong, and X. C. Yuan. Composite-microlens-array-enabled microfluidic sorting. *Applied Physics Letters*, 89, (14): 141108, **2006**.

[80] F. Merenda, J. Rohner, J. M. Fournier, and R. P. SalathâĹŽÂĬ. Miniaturized high-NA focusing-mirror multiple optical tweezers. *Optics Express*, 15, (10): 6075–6086, **2007**. 35

[81] M. P. MacDonald, L. Paterson, W. Sibbett, K. Dholakia, and P. E. Bryant. Trapping and manipulation of low-index particles in a two-dimensional interferometric optical trap. *Optics Letters*, 26, (12): 863–865, **2001**. 35

[82] M. P. MacDonald, G. C. Spalding, and K. Dholakia. Microfluidic sorting in an optical lattice. *Nature*, 426, (6965): 421–4, **2003**.

[83] V. Garces-Chavez, K. Dholakia, and G. C. Spalding. Extended-area optically induced organization of microparticles on a surface. *Applied Physics Letters*, 86, (3): 031106, **2005**. 35

[84] D. Boiron, A. Michaud, J. M. Fournier, L. Simard, M. Sprenger, G. Grynberg, and C. Salomon. Cold and dense cesium clouds in far-detuned dipole traps. *Physical Review A*, 57, (6): 4106–4109, **1998**. 35

[85] E. Schonbrun, R. Piestun, P. Jordan, J. Cooper, K. Wulff, J. Courtial, and M. Padgett. 3D interferometric optical tweezers using a single spatial light modulator. *Optics Express*, 13, (10): 3777–3786, **2005**. 35

[86] P. C. Mogensen and J. Glueckstad. Dynamic array generation and pattern formation for optical tweezers. *Optics Communications*, 175, (1-3): 75–81, **2000**. 35

[87] R. L. Eriksen, P. C. Mogensen, and J. Glueckstad. Multiple-beam optical tweezers generated by the generalized phase-contrast method. *Optics Letters*, 27, (4): 267–269, **2002**. 35

[88] E. R. Dufresne and D. G. Grier. Optical tweezer arrays and optical substrates created with diffractive optics. *Review of Scientific Instruments*, 69: 1974, **1998**. 35

[89] David G. Grier. A revolution in optical manipulation. *Nature Photonics*, 424, (6950): 810–816, **2003**. 35

[90] A. Rosenhahn. Analytik Bilder aus der Tiefe: Digitale In-line-Holographie. *Nachrichten aus der Chemie*, 56, (1): 55, **2008**. 36

[91] L. B. Lesem, P. M. Hirsch, and J. A. Jordan Jr. The kinoform: a new wavefront reconstruction device. *IBM J. Res. Dev*, 13, (1): 150âĂŽĂĎĂň155, **1969**. 37

[92] R. W. Gerchberg and W. O. Saxton. A practical algorithm for the determination of phase from image and diffraction plane pictures. *SPIE Milestone Series*, 93: 306–306, **1994**. 37

[93] C. H. J. Schmitz, J. P. Spatz, and J. E. Curtis. High-precision steering of multiple holographic optical tweezers. *Optics Express*, 13: 8678–8685, **2005**. 42, 44

[94] H. Lodish, A. Berk, S. L. Zipursky, P. Matsudaira, D. Baltimore, and J. Darnell. *Molecular cell biology*. W,H. Freeman, **2000**. 45

[95] J. Howard and R. L. Clark. Mechanics of motor proteins and the cytoskeleton. *Applied Mechanics Reviews*, 55, (2): B39, **2002**. 45

[96] P. A. Janmey. The cytoskeleton and cell signaling: Component localization and mechanical coupling. *Physiology Reviews*, 78, (3): 763–781, **1998**.

[97] A.J. Engler, S. Sen, H. L. Sweeney, and D. E. Discher. Matrix elasticity directs stem cell lineage specification. *Cell*, 126, (4): 677–689, **2006**. 45

[98] T. Kreis and R. Vale. Guidebook to the cytoskeletal and motor proteins. *Guide book series, Oxford Press*, **2004**. 46, 47

[99] F. Gittes, B. Mickey, J. Nettleton, and J. Howard. Flexural rigidity of microtubules and actin filaments measured from thermal fluctuations in shape. *Journal of Cell Biology*, 120, (4): 923–934, **1993**. 46, 47

[100] F. Pampaloni, G. Lattanzi, A. Jonas, T. Surrey, E. Frey, and E.-L. Florin. Thermal fluctuations of grafted microtubules provide evidence of a length-dependent persistence length. *Proceedings of the National Academy of Sciences*, 103, (27): 10248–10253, **2006**. 46

[101] C.P. Brangwynne, F. C. MacKintosh, S. Kumar, N. A. Geisse, J. Talbot, L. Mahadevan, K. Parker, D. E. Ingber, and D. A. Weitz. Microtubules can bear enhanced compressive loads in living cells because of lateral reinforcement. *Journal of Cell Biology*, 173, (5): 733–741, **2006**. 46

[102] D. E. Ingber. Tensegrity: The architectural basis of cellular mechanotransduction. *Annual Reviews in Physiology*, 59, (1): 575–599, **1997**. 46, 47

[103] H. Baribault, J. Price, K. Miyai, and R. G. Oshima. Mid-gestational lethality in mice lacking keratin 8. *Genes & Development*, 7, (7a): 1191, **1993**. 46

[104] E. Fuchs and K. Weber. Intermediate filaments: Structure, dynamics, function and disease. *Annual Reviews in Biochemistry*, 63, (1): 345–382, **1994**. 46

[105] R. D. Goldman, S. Khuon, Y. H. Chou, P. Opal, and P. M. Steinert. The function of intermediate filaments in cell shape and cytoskeletal integrity. *Journal of Cell Biololgy*, 134, (4): 971–983, **1996**. 46

[106] T. D. Pollard and G. G. Borisy. Cellular motility driven by assembly and disassembly of actin filaments. *Cell*, 112, (4): 453–65, **2003**. 47, 95

[107] J. W. Sanger. Changing patterns of actin localization during cell division. *Proceedings of the National Academy of Sciences*, 72, (5): 1913–1916, **1975**.

[108] G. Eitzen. Actin remodeling to facilitate membrane fusion. *Biochimica et Biophysica Acta (BBA) - Molecular Cell Research*, 1641, (2-3): 175–181, **2003**. 52

[109] B. Qualmann, M. M. Kessels, and R. B. Kelly. Molecular links between endocytosis and the actin cytoskeleton. *Journal of Cell Biololgy*, 150, (5): F111–6, **2000**. 95

[110] L. Miao, O. Vanderlinde, M. Stewart, and T. M. Roberts. Retraction in amoeboid cell motility powered by cytoskeletal dynamics. *Science*, 302, (5649): 1405–1407, **2003**. 60

[111] J. Hanson and H. E. Huxley. Structural basis of the cross-striations in muscle. *Nature*, 172, (4377): 530–532, **1953**. 50

[112] I. Rayment, H. M. Holden, M. Whittaker, C. B. Yohn, M. Lorenz, K. C. Holmes, and R. A. Milligan. Structure of the actin-myosin complex and its implications for muscle contraction. *Science*, 261, (5117): 58, **1993**. 47

[113] J. A. Traas, J. H. Doonan, D. J. Rawlins, P. J. Shaw, J. Watts, and C. W. Lloyd. An actin network is present in the cytoplasm throughout the cell cycle of carrot cells and associates with the dividing nucleus. *Journal of Cell Biology*, 105, (1): 387–395, **1987**. 47, 52

[114] A. R. Bausch, F. Ziemann, A. A. Boulbitch, K. Jacobson, and E. Sackmann. Local measurements of viscoelastic parameters of adherent cell surfaces by magnetic bead microrheometry. *Biophysical Journal*, 75, (4): 2038–2049, **1998**.

[115] M. L. Gardel, F. Nakamura, J. Hartwig, J. C. Crocker, T. P. Stossel, and D. A. Weitz. Stress-dependent elasticity of composite actin networks as a model for cell behavior. *Physical Review Letters*, 96, (8): 88102, **2006**. 95

[116] Thorsten Lang, Irene Wacker, Ilse Wunderlich, Alexander Rohrbach, Gunter Giese, Thierry Soldati, and Wolfhard Almers. Role of Actin Cortex in the Subplasmalemmal Transport of Secretory Granules in PC-12 Cells. *Biophysical Journal*, 78, (6): 2863–2877, **2000**. 47

[117] E. Evans, K. Ritchie, and R. Merkel. Sensitive force technique to probe molecular adhesion and structural linkages at biological interfaces. *Biophysical Journal*, 68, (6): 2580–2587, Jun 1995. 47, 53

[118] C. Revenu, R. Athman, S. Robine, and D. Louvard. The co-workers of actin filaments: from cell structures to signals. *Nature Reviews Molecular Cell Biology*, 5: 635–646, **2004**. 47

[119] A. Bremer. The structural basis for the intrinsic disorder of the actin filament: the" lateral slipping" model. *The Journal of Cell Biology*, 115, (3): 689–703, **1991**. 47

[120] K. C. Holmes. Solving the structures of macromolecular complexes. *Structure*, 2, (7): 589–593, **1994**.

[121] P. A. Janmey, J. X. Tang, and C. F. Schmidt. Actin filaments. *Supramolecular Assemblies*, **2001**. 95

[122] T. Oda, K. Makino, I. Yamashita, K. Namba, and Y. Maeda. The helical parameters of F-actin precisely determined from X-ray fiber diffraction of well-oriented sols. *Results Probl Cell Differ*, 32: 43–58, **2001**. 47

[123] A. Ott, M. Magnasco, A. Simon, and A. Libchaber. Measurement of the persistence length of polymerized actin using fluorescence microscopy. *Physical Review E*, 48, (3): 1642–1645, **1993**. 47

[124] J. Xu, A. Palmer, and D. Wirtz. Rheology and microrheology of semiflexible polymer solutions: Actin filament networks. *Macromolecules*, 31: 6486–6492, **1998**. 47

[125] S. Schmitz, M. Grainger, S. Howell, L. J. Calder, M. Gaeb, J. C. Pinder, A. A. Holder, and C. Veigel. Malaria parasite actin filaments are very short. *Journal of Molecular Biology*, 349, (1): 113–125, **2005**. 48, 61

[126] C. Frieden. Actin and tubulin polymerization: The use of kinetic methods to determine mechanism. *Annual Reviews in Biophysics and Biophysical Chemistry*, 14, (1): 189–210, **1985**. 48

[127] T. D. Pollard. Rate constants for the reactions of atp-and adp-actin with the ends of actin filaments. *Journal of Cell Biology*, 103, (6): 2747–2754, **1986**. 48

[128] T. D. Pollard and J. A. Cooper. Actin and actin-binding proteins. a critical evaluation of mechanisms and functions. *Annual Reviews in Biochemistry*, 55, (1): 987–1035, **1986**. 48

[129] A. Weeds and S. Maciver. F-actincapping proteins. *Current Opinion in Cell Biology*, 5, (1): 63–69, **1993**.

[130] L. D. Burtnick, E. K. Koepf, J. Grimes, E. Y. Jones, D. I. Stuart, P. J. McLaughlin, and R. C. Robinson. The crystal structure of plasma gelsolin: Implications for actin severing, capping, and nucleation. *Cell*, 90: 661–670, **1997**.

[131] J. Condeelis. How is actin polymerization nucleated in vivo? *Trends in Cell Biology*, 11, (7): 288–293, **2001**. 48

[132] L. M. Griffith and T. D. Pollard. Cross-linking of actin filament networks by self-association and actin-binding macromolecules. *Journal of Biological Chemistry*, 257, (15): 9135–9142, **1982**. 49

[133] Y. Yang, J. Dowling, Q. C. Yu, P. Kouklis, D. W. Cleveland, and E. Fuchs. An essential cytoskeletal linker protein connecting actin microfilaments to intermediate filaments. *Cell*, 86, (4): 655–65, **1996**.

[134] J. Taunton, B. A. Rowning, M. L. Coughlin, M. Wu, R. T. Moon, T. J. Mitchison, and C. A. Larabell. Actin-dependent propulsion of endosomes and lysosomes by recruitment of N-WASP. *Journal of Cell Biology*, 148, (3): 519–530, **2000**. 49

[135] J. Howard. Molecular motors: structural adaptations to cellular functions. *Nature*, 389: 561–567, **1997**. 49

[136] C. H. J. Schmitz. Entwicklung eines optomechanischen Mikrolabors zur Generierung und Untersuchung biomimetischer Proteinnetzwerke. *Dissertation*, **2005**. 50

[137] W. R. Middlebrook. Individuality of the meromyosins. *Science*, 130, (3376): 621–622, **1959**. 51

[138] S. Lowey, H. S. Slayter, A. G. Weeds, and H. Baker. Substructure of the myosin molecule. i. subfragments of myosin by enzymic degradation. *Journal of Molecular Biology*, 42, (1): 1–29, **1969**. 51

[139] R. L. Meeusen and W. Z. Cande. N-ethylmaleimide-modified heavy meromyosin. a probe for actomyosin interactions. *Journal of Cell Biology*, 82, (1): 57–65, **1979**. 51

[140] S. Ebashi and F. Ebashi. Alpha-Actinin, a New Structural Protein from Striated Muscle: 1. Preparation and Action on Actomyosin-ATP Interaction. *Journal of Biochemistry*, 58, (1): 7, **1965**. 51

[141] M. Edlund, M. A. Lotano, and C. A. Otey. Dynamics of alpha-actinin in focal adhesions and stress fibers visualized with alpha-actinin-green fluorescent protein. *Cell Motility and the Cytoskeleton*, 48, (3): 190–200, **2001**. 51

[142] D. H. Wachsstock, W. H. Schwartz, and T. D. Pollard. Affinity of alpha-actinin for actin determines the structure and mechanical properties of actin filament gels. *Biophysical Journal*, 65, (1): 205–214, **1993**. 51

[143] M. Tempel, G. Isenberg, and E. Sackmann. Temperature-induced sol-gel transition and microgel formation in alpha -actinin cross-linked actin networks: A rheological study. *Physical Review E*, 54, (2): 1802, **1996**.

[144] O. Pelletier, E. Pokidysheva, L. S. Hirst, N. Bouxsein, Y. Li, and C. R. Safinya. Structure of actin cross-linked with alpha-actinin: A network of bundles. *Physical Review Letters*, 91, (14): 148102, **2003**. 51, 52, 111

[145] P. Defilippi, C. Olivo, M. Venturino, L. Dolce, L. Silengo, and G. Tarone. Actin cytoskeleton organization in response to integrin-mediated adhesion. *Microscopy Research and Technique*, 47, (1): 67–78, **1999**. 52

[146] A. S. Sechi and J. Wehland. The actin cytoskeleton and plasma membrane connection: Ptdins(4,5)p(2) influences cytoskeletal protein activity at the plasma membrane. *Journal of Cell Science*, 113, (21): 3685–3695, **2000**. 52

[147] P. G. de Gennes. *Scaling Concepts in Polymer Physics*. Cornell University Press, **1979**. 52

[148] O. Lieleg, M. Claessens, C. Heussinger, E. Frey, and A. R. Bausch. Mechanics of bundled semiflexible polymer networks. *Physical Review Letters*, 99, (8): 88102, **2007**. 52

[149] A. Bretscher and K. Weber. Localization of actin and microfilament-associated proteins in the microvilli and terminal web of the intestinal brush border by immunofluorescence microscopy. *Journal of Cell Biology*, 79, (3): 839–845, **1978**. 53

[150] J. R. Bartles. Parallel actin bundles and their multiple actin-bundling proteins. *Current Opinion in Cell Biology*, 12, (1): 72–78, **2000**. 53, 111

[151] R. Pelham and F. Chang. Actin dynamics in the contractile ring during cytokinesis in fission yeast. *Nature*, 419, (6902): 82–86, **2002**. 53, 95

[152] J. Uhde, M. Keller, E. Sackmann, A. Parmeggiani, and E. Frey. Internal motility in stiffening actin-myosin networks. *Physical Review Letters*, 93, (26): 268101, **2004**. 54, 111

[153] M. Aregawi, R. Cibulskis, and M. Otten. *World Malaria Report 2008*. World Health Organization, **2008**. 55

[154] M. Chan. *The World Health Report 2008: Primary Health Care Now More Than Ever*. World Health Organization, **2008**. 55, 56

[155] P. Martens, R. S. Kovats, S. Nijhof, P. De Vries, M. T. J. Livermore, D. J. Bradley, J. Cox, and A. J. McMichael. Climate change and future populations at risk of malaria. *Global Environmental Change, Part A: Human and Policy Dimensions*, 9, **1999**. 55

[156] D. J. Rogers and S. E. Randolph. The global spread of malaria in a future, warmer world, **2000**.

[157] K. G. Kuhn, D. H. Campbell-Lendrum, B. Armstrong, and C. R. Davies. Malaria in britain: past, present, and future. *Proceedings of the National Academy of Sciences*, 100, (17): 9997–10001, **2003**. 55

[158] L. D. Sibley. Intracellular parasite invasion strategies. *Science*, 304, (5668): 248–253, **2004**. 55

[159] J. Baum, A. T. Papenfuss, B. Baum, T. P. Speed, and A. F. Cowman. Regulation of apicomplexan actin-based motility. *Nature Reviews Microbiology*, 4, (8): 621–628, **2006**. 55, 60, 61

[160] N. S. Morrissette and L. D. Sibley. Cytoskeleton of apicomplexan parasites. *Microbiology and Molecular Biology Reviews*, 66, (1): 21–38, **2002**. 55

[161] J. L. Jones, A. Lopez, M. Wilson, J. Schulkin, and R. Gibbs. Congenital toxoplasmosis: A review. *Obstetrical & Gynecological Survey*, 56, (5): 296, **2001**.

[162] L. M. Weiss. Babesiosis in humans: a treatment review. *Expert Opinion on Pharmacotherapy*, 3, (8): 1109–1115, **2002**. 55

[163] D. A. Joy, X. Feng, J. Mu, T. Furuya, K. Chotivanich, A. U. Krettli, M. Ho, A. Wang, Nicholas J. White, E. Suh, P. Beerli, and X. Su. Early origin and recent expansion of plasmodium falciparum. *Science*, 300, (5617): 318–321, **2003**. 56

[164] Tonse N. K. Raju. Hot brains: Manipulating body heat to save the brain. *Pediatrics*, 117, (2): e320–321, **2006**. 56

[165] V. Ebersbach and A. Siekmann. *Anekdotenüber Goethe und Schiller*. Medienservice Mathias Karge, **2005**. 56

[166] R. W. Snow, C. A. Guerra, A. M. Noor, H.Y. Myint, and S. I. Hay. The global distribution of clinical episodes of plasmodium falciparum malaria. *Nature*, 434, (7030): 214–217, **2005**. 56

[167] C. G. Meyer, F. Marks, and J. May. Editorial: Gin tonic revisited. *Tropical Medicine and International Health*, 9, (12): 1239–1240, **2004**. 56

[168] R. Tuteja. Malaria-an overview. *FEBS Journal*, 274, (18): 4670–4679, **2007**. 57

[169] T. E. Wellems and C. V. Plowe. Chloroquine-resistant malaria. *Journal of Infectious Diseases*, 184, (6): 770–776, **2001**. 57

[170] M. M. Mota and A. Rodriguez. Invasion of mammalian host cells by plasmodium sporozoites. *Bioessays*, 24, (2), **2002**. 57

[171] J. P. Vanderberg. Studies on the motility of plasmodium sporozoitese. *Journal of Eukaryotic Microbiology*, 21, (4): 527–537, **1974**. 57, 61, 92

[172] P. Sinnis and F. Zavala. The skin stage of malaria infection: biology and relevance to the malaria vaccine effort. *Future Microbiology*, 3, (3): 275–278, **2008**. 57

[173] M. M. Mota, J. C. R. Hafalla, and A. Rodriguez. Migration through host cells activates plasmodium sporozoites for infection. *Nature Medizine*, 8, (11): 1318–1322, **2002**. 57, 141

[174] Miguel Prudencio, Ana Rodriguez, and Maria M. Mota. The silent path to thousands of merozoites: the plasmodium liver stage. *Nature Reviews Microbiology*, 4, (11): 849–856, **2006**.

[175] A. Sturm, R. Amino, C. van de Sand, T. Regen, S. Retzlaff, A. Rennenberg, A. Krueger, J.-M. Pollok, R. Menard, and V. T. Heussler. Manipulation of host hepatocytes by the malaria parasite for delivery into liver sinusoids. *Science*, 313, (5791): 1287–1290, **2006**. 57

[176] M. Aikawa, L. H. Miller, J. Johnson, and J. Rabbege. Erythrocyte entry by malarial parasites. a moving junction between erythrocyte and parasite. *Journal of Cell Biology*, 77, (1): 72–82, **1978**. 59

[177] K. Wengelnik, R. Spaccapelo, S. Naitza, K. J. H. Robson, C. J. Janse, F. Bistoni, A. P. Waters, and A. Crisanti. The A-domain and the thrombospondin-related motif of Plasmodium falciparum TRAP are implicated in the invasion

process of mosquito salivary glands. *EMBO Journal*, 18: 5195–5204, **1999**. 59, 141

[178] A. K. Ghosh, M. Devenport, D. Jethwaney, D. E. Kalume, A. Pandey, V. E. Anderson, A. A. Sultan, N. Kumar, and M. Jacobs-Lorena. Malaria parasite invasion of the mosquito salivary gland requires interaction between the plasmodium trap and the anopheles saglin proteins. *PLoS Pathogens*, 5, (1), **2009**. 59

[179] F. Frischknecht, P. Baldacci, B. Martin, C. Zimmer, S. Thiberge, J. C. Olivo-Marin, S. L. Shorte, and R. Menard. Imaging movement of malaria parasites during transmission by anopheles mosquitoes. *Cellular Microbiology*, 6, (7): 687–694, **2004**. 59

[180] L. Florens, M. P. Washburn, J. D. Raine, R. M. Anthony, M. Grainger, J. D. Haynes, J. K. Moch, N. Muster, J. B. Sacci, and D. L. Tabb. A proteomic view of the plasmodium falciparum life cycle. *Nature*, 419: 520–526, **2002**. 60

[181] L. A. Cameron, P. A. Giardini, F. S. Soo, and J. A. Theriot. Secrets of actin-based motility revealed by a bacterial pathogen. *Nature Reviews Molecular Cell Biology*, 1, (2): 110–119, **2000**. 60

[182] C. A. King. Cell motility of sporozoan protozoa. *Parasitology Today*, 4, (11): 315–319, **1988**. 60

[183] D. G. Russell and R. E. Sinden. The role of the cytoskeleton in the motility of coccidian sporozoites. *Journal of Cell Science*, 50, (1): 345–359, **1981**. 61

[184] J. R. Forney, D. K. Vaughan, S. Yang, and M. C. Healey. Actin-dependent motility in cryptosporidium parvum sporozoites. *Journal of Parasitology*, 84, (5): 908–13, **1998**. 61

[185] J. Baum, D. Richard, J. Healer, M. Rug, Z. Krnajski, T. W. Gilberger, J. L. Green, A. A. Holder, and A. F. Cowman. A conserved molecular motor drives cell invasion and gliding motility across malaria life cycle stages and other apicomplexan parasites. *Journal of Biological Chemistry*, 281, (8): 5197, **2006**. 61, 62

[186] C. Lacroix and R. Menard. TRAP-like protein of Plasmodium sporozoites: linking gliding motility to host-cell traversal. *Trends in Parasitology*, **2008**. 61, 62

[187] K. Heiss, S. Nie, H.and Kumar, T. M. Daly, L. W. Bergman, and K. Matuschewski. Functional characterization of a redundant plasmodium TRAP family invasin, TRAP-like protein, by aldolase binding and a genetic complementation test. *Eukaryotic Cell*, 7, (6): 1062–1070, **2008**. 61

[188] A. Keeley and D. Soldati. The glideosome: a molecular machine powering motility and host-cell invasion by apicomplexa. *Trends in Cell Biology*, 14, (10): 528–532, **2004**. 61

[189] C. K. Moreira, T. J. Templeton, C. Lavazec, R. E. Hayward, C. V. Hobbs, H. Kroeze, C. J. Janse, A. P. Waters, P. Sinnis, and A. Coppi. The plasmodium TRAP/MIC2 family member, TRAP-Like Protein (TLP), is involved in tissue traversal by sporozoites. *Cellular Microbiology*, 10, (0): 1505–1516, **2008**. 61, 62

[190] K. Kaiser, K. Matuschewski, N. Camargo, J. Ross, and S. H. I. Kappe. Differential transcriptome profiling identifies plasmodium genes encoding pre-erythrocytic stage-specific proteins. *Molecular Microbiology*, 51, (5): 1221–1232, **2004**. 61, 141

[191] J. T. Dessens, A. L. Beetsma, G. Dimopoulos, K. Wengelnik, A. Crisanti, F. C. Kafatos, and R. E. Sinden. Ctrp is essential for mosquito infection by malaria ookinetes. *The EMBO Journal*, 18: 6221–6227, **1999**. 62

[192] V. Nussenzweig and R. Menard. Analysis of a malaria sporozoite protein family required for gliding motility and cell invasion. *Trends in Microbiology*, 8, (3): 94–96, **2000**.

[193] J. L. Green, L. Hinds, M. Grainger, E. Knuepfer, and A. A. Holder. Plasmodium thrombospondin related apical merozoite protein (ptramp) is shed from the surface of merozoites by pfsub2 upon invasion of erythrocytes. *Molecular & Biochemical Parasitology*, 150, (1): 114–117, **2006**. 62

[194] R. Amino, R. Menard, and F. Frischknecht. In vivo imaging of malaria parasitesâĂŽ recent advances and future directions. *Current Opinion in Microbiology*, 8, (4): 407–414, **2005**. 61

[195] J. C. Crocker and D. G. Grier. Methods of digital video microscopy for colloidal studies. *Journal of Colloid and Interface Science*, 179: 298–310, **1996**. 75

[196] W. H. Roos, A. Roth, J. Konle, H. Presting, E. Sackmann, and J. P. Spatz. Freely suspended actin cortex models on arrays of microfabricated pillars. *ChemPhysChem*, 4, (8): 872–7, **2003**. 77, 95, 131

[197] D. C. Duffy, J. C. McDonald, O. J. A. Schueller, and G. M. Whitesides. Rapid Prototyping of Microfluidic Systems in Poly(dimethylsiloxane). *Analytical Chemistry*, 70: 4974–4984, **1998**. 77

[198] Y. N. Xia and G. M. Whitesides. Softlithography. *Angewandte Chemie Int. Ed. Engl.*, 37: 551–577, **1998**. 77

[199] C. H. J. Schmitz, A. C. Rowat, S. Koster, and D. A. Weitz. Dropspots: a picoliter array in a microfluidic device. *Lab on a Chip*, 9, (1): 44–49, **2009**. 81

[200] M. Morra, E. Occhiello, R. Marola, F. Garbassi, P. Humphrey, and D. Johnson. On the aging of oxygen plasma-treated polydimethylsiloxane surfaces. *Journal of Colloid and Interface Science*, 137, (1): 11âĂŽĂĎĂň24, **1990**. 81

[201] M. K. Chaudhury and G. M. Whitesides. Direct measurement of interfacial interactions between semispherical lenses and flat sheets of poly (dimethylsiloxane) and their chemical derivatives. *Langmuir*, 7, (5): 1013–1025, **1991**.

[202] M. K. Chaudhury and G. M. Whitesides. Correlation between surface free energy and surface constitution. *Science*, 255, (5049): 1230–1232, **1992**. 81

[203] J. W. Park Lin, M. Protein solubility in pacific whiting affected by proteolysis during storage. *Journal of Food Science*, 61, (3): 536–539, **1996**. 85

[204] A. Coffey, van den Burg R., and Veltman T. A. Characteristics of the biologically active 35-kDa metalloprotease virulence factor from Listeria monocytogenes. *Journal of Applied Microbiology*, 88, (1): 132–141, **2000**.

[205] A. V. Morozova, I. N. Skovorodkin, S. Yu Khaitlina, and A. Yu Malinin. Bacterial protease ecp32 specifically hydrolyzing actin and its effect on cytoskeleton in vivo. *Biochemistry (Moscow)*, 66, (1): 83–90, **2001**. 85

[206] J. X. Tang, P. A. Janmey, T. P. Stossel, and T. Ito. Thiol oxidation of actin produces dimers that enhance the elasticity of the F-Actin network. *Biophysical Journal*, 76, (4): 2208–2215, **1999**. 85

[207] S. Ishiwata. Freezing of actin. reversible oxidation of a sulfhydryl group and structural change. *Journal of Biochemistry*, 80, (3): 595, **1976**. 85

[208] J. D. Pardee and J. A. Spudich. Purification of muscle actin. *Methods Enzymol*, 85 Pt B: 164–81, **1982**. 85

[209] S. MacLean-Fletcher and T. D. Pollard. Identification of a factor in conventional muscle actin preparations which inhibits actin filament self-association. *Biochem Biophys Res Commun*, 96, (1): 18–27, **1980**. 85

[210] D. Bray and J. G. White. Cortical flow in animal cells. *Science*, 239, (4842): 883–888, **1988**. 95

[211] B. Geiger. Membrane-cytoskeleton interaction. *Biochimica et Biophysica Acta (BBA) - Reviews on Biomembranes*, 737, (3-4): 305–341, **1983**. 95

[212] J. M. Vasiliev. Polarization of pseudopodial activities: cytoskeletal mechanisms. *Journal of Cell Science*, 98, (1): 1–4, **1991**. 95

[213] N. Q. Balaban, U. S. Schwarz, D. Riveline, P. Goichberg, G. Tzur, I. Sabanay, D. Mahalu, S. Safran, A. Bershadsky, L. Addadi, and B. Geiger. Force and focal adhesion assembly: a close relationship studied using elastic micropatterned substrates. *Nature Cell Biology*, 3, (5): 466–472, **2001**. 95

[214] F. C. MacKintosh, J. KâLŽÂǧs, and P. A. Janmey. Elasticity of semiflexible biopolymer networks. *Physical Review Letters*, 75, (24): 4425, **1995**. 95

[215] S. Kaufmann, J. Kas, W. H. Goldmane, and E. Sackmann. Talin anchors and nucleates actin filaments a direct demonstration. *FEBS Letters*, 314: 203–205, **1992**. 95

[216] L. Le Goff, O. Hallatschek, E. Frey, and F. Amblard. Tracer studies on F-Actin fluctuations. *Physical Review Letters*, 89, (25): 258101, **2002**. 95

[217] F. Amblard, A. C. Maggs, B. Yurke, A. N. Pargellis, and S. Leibler. Subdiffusion and anomalous local viscoelasticity in actin networks. *Physical Review Letters*, 77, (21): 4470–4473, **1996**.

[218] F. Gittes, B. Schnurr, P. D. Olmsted, F. C. MacKintosh, and C. F. Schmidt. Microscopic viscoelasticity: Shear moduli of soft materials determined from thermal fluctuations. *Physical Review Letters*, 79, (17): 3286, **1997**. 95

[219] P. A. Janmey, S. Hvidt, J. Kas, D. Lerche, A. Maggs, E. Sackmann, M. Schliwa, and T. P. Stossel. The mechanical properties of actin gels. elastic modulus and filament motions. *Journal of Biological Chemistry*, 269, (51): 32503–32513, **1994**. 95

[220] M. L. Gardel, J. H. Shin, F. C. MacKintosh, L. Mahadevan, P. Matsudaira, and D. A. Weitz. Elastic behavior of cross-linked and bundled actin networks. *Science*, 304, (5675): 1301–1305, **2004**.

[221] P. M. Bendix, G. H. Koenderink, D. Cuvelier, Z. Dogic, B. N. Koeleman, W. M. Brieher, C. M. Field, L. Mahadevan, and D. A. Weitz. A quantitative analysis of contractility in active cytoskeletal protein networks. *Biophysical Journal*, 94, (8): 3126–36, **2008**. 95

[222] K. A. Taylor and D. W. Taylor. Structural studies of cytoskeletal protein arrays formed on lipid monolayers. *Journal of Structural Biology*, 128, (1): 75–81, **1999**. 95

[223] C. H. Schmitz, J. Curtis, and J. P. Spatz. Constructing and probing biomimetic models of the actin cortex with holographic optical tweezers. *Proceedings of SPIE*, 5514: 446–454, **2004**. 95

[224] F. Gittes and C. F. Schmidt. Signals and noise in micromechanical measurements. *Methods Cell Biol*, 55: 129–56, **1998**. 102

[225] K. Berg-Sorensen and H. Flyvbjerg. Power spectrum analysis for optical tweezers. *Review of Scientific Instruments*, 75: 594–612, **2004**. 104

[226] X. Liu and G. H. Pollack. Mechanics of F-actin characterized with microfabricated cantilevers. *Biophysical Journal*, 83, (5): 2705–15, **2002**. 106

[227] J. X. Tang and P. A. Janmey. The polyelectrolyte nature of F-actin and the mechanism of actin bundle formation. *Journal of Biological Chemistry*, 271, (15): 8556–63, **1996**. 107, 111, 119

[228] H. Qian, M. P. Sheetz, and E. L. Elson. Single particle tracking. analysis of diffusion and flow in two-dimensional systems. *Biophysical Journal*, 60, (4): 910–21, **1991**. 108, 116

[229] Y. Luan, O. Lieleg, B. Wagner, and A. R. Bausch. Micro- and macrorheological properties of isotropically cross-linked actin networks. *Biophysical Journal*, 94, (2): 688–693, **2008**. 108

[230] R. Furukawa and M. Fechheimer. The structure, function, and assembly of actin filament bundles. *International review of cytology*, 175: 29–90, **1997**. 111

[231] G. S. Manning. The molecular theory of polyelectrolyte solutions with applications to the electrostatic properties of polynucleotides. *Quarterly Reviews of Biophysics*, 11, (2): 179–246, **1978**.

[232] K. Kruse and F. Juelicher. Self-organization and mechanical properties of active filament bundles. *Physical Review E*, 67, (5): 051913, **2003**. 111

[233] C. W. Jones, J. C. Wang, R. W. Briehl, and M. S. Turner. Measuring forces between protein fibers by microscopy. *Biophysical Journal*, 88, (4): 2433–2441, **2005**. 111

[234] J. Kierfeld, P. Gutjahr, T. Kuhne, P. Kraikivski, and R. Lipowsky. Buckling, bundling, and pattern formation: From semi-flexible polymers to assemblies of interacting filaments. *Journal of Computational and Theoretical Nanonscience*, 3, (6): 898, **2006**. 111, 118, 123

[235] S. Chandrasekhar. Stochastic problems in physics and astronomy. *Reviews of Modern Physics*, 15, (1): 1–89, **1943**. 116

[236] J. Kierfeld. *Personal Communication*. 119

[237] D. C. Rau and V. A. Parsegian. Direct measurement of the intermolecular forces between counterion-condensed DNA double helices. Evidence for long range attractive hydration forces. *Biophysical Journal*, 61, (1): 246–259, **1992**. 119

[238] T. T. Nguyen, I. Rouzina, and B. I. Shklovskii. Reentrant condensation of DNA induced by multivalent counterions. *Journal of Chemical Physics*, 112: 2562, **2000**.

[239] B. Y. Ha and Andrea J. Liu. Counterion-mediated attraction between two like-charged rods. *Physical Review Letters*, 79, (7): 1289, **1997**. 119

[240] J. X. Tang, P. T. Szymanski, P.A. Janmey, and T. Tao. Electrostatic effects of smooth muscle calponin on actin assembly. *European Journal of Biochemistry*, 247, (1): 432–440, **1997**. 120

[241] F. Wohnsland, A. A.P. Schmitz, M. O. Steinmetz, U. Aebi, and G. Vergeres. Interaction between actin and the effector peptide of MARCKS-related protein. identification of functional amino acid segments. *Journal of Biological Chemistry*, 275, (27): 20873–20879, **2000**. 120

[242] J. Kierfeld, T. Kühne, and R. Lipowsky. Discontinuous unbinding transitions of filament bundles. *Physical Review Letters*, 95, (3): 38102, **2005**. 121

[243] C. Friedsam, A. D. C. Becares, U. Jonas, M. Seitz, and H. E. Gaub. Adsorption of polyacrylic acid on self-assembled monolayers investigated by single-molecule force spectroscopy. *New Journal of Physics*, 6, (1): 9, **2004**. 123

[244] P.K. Mishra, S. Kumar, and Y. Singh. Force-induced desorption of a linear polymer chain adsorbed on an attractive surface. *Europhys. Lett*, 69, (1): 102, **2005**. 123

[245] J. E. Curtis, C. H. J. Schmitz, and J. P. Spatz. Symmetry dependence of holograms for optical trapping. *Optics Letters*, 30, (16): 2086–2088, **2005**. 127

[246] M. Polin, K. Ladayac, S. H. Lee, Y. Roichman, and D. G. Grier. Optimized holographic optical traps. *Optics Express*, 13: 5831–5845, **2005**. 128

[247] G. Milewski, D. Engstroem, and J. Bengtsson. Diffractive optical elements designed for highly precise far-field generation in the presence of artifacts typical for pixelated spatial light modulators. *Applied Optics*, 46, (1): 95–105, **2007**. 128

[248] C. Mohrdieck, A. Wanner, W. Roos, A. Roth, E. Sackmann, J. Spatz, and E. Arzt. A theoretical description of elastic pillar substrates in biophysical experiments. *ChemPhysChem*, 6, (8): 1492–1498, **2005**. 131

[249] C. Mohrdieck, F. Dalmas, E. Arzt, R. Tharmann, M. Claessens, A. Bausch, A. Roth, E. Sackmann, C. Schmitz, J. Curtis, W. Roos, S. Schulz, K. Uhrig, and J. Spatz. Biomimetic models of the actin cytoskeleton. *Small*, 3, (6): 1015–1022, **2007**. 131

[250] E. Evans. Probing the relation between force kifetime and chemistry in single molecular bonds. *Annual Reviews in Biophysics and Biomolecular Structure*, 30, (1): 105–128, **2001**. 137

[251] O. K. Dudko, G. Hummer, and A. Szabo. Intrinsic rates and activation free energies from single-molecule pulling experiments. *Physical Review Letters*, 96, (10): 108101, **2006**. 137

[252] E. Evans and K. Ritchie. Dynamic strength of molecular adhesion bonds. *Biophysical Journal*, 72, (4): 1541–1555, Apr 1997. 138

[253] P. M. Williams. Analytical descriptions of dynamic force spectroscopy: behaviour of multiple connections. *Analytica Chimica Acta*, 479, (1): 107–115, **2003**. 138

[254] J. Kierfeld. Force-induced desorption and unzipping of semiflexible polymers. *Physical Review Letters*, 97, (5): 58302, **2006**. 140

[255] Belinda J. Morahan, Lina Wang, and Ross L. Coppel. No TRAP, no invasion. *Trends in Parasitology*, In Press, Corrected Proof, **2009**. 141

[256] P. Baldacci and R. Menard. The elusive malaria sporozoite in the mammalian host. *Molecular Microbiology*, 54, (2): 298–306, **2004**. 141

[257] M. Steinbuechel and K. Matuschewski. Role for the plasmodium sporozoite-specific transmembrane protein s6 in parasite motility and efficient malaria transmission. *Cellular Microbiology*, 11, (2): 279–288, **2009**. 141

[258] Y. Liu, G. J. Sonek, M. W. Berns, and B. J. Tromberg. Physiological monitoring of optically trapped cells: assessing the effects of confinement by 1064 nm laser tweezers using microfluorometry. *Biophysical Journal*, 71, (4): 2158–2167, **1996**. 142

[259] K. C. Neuman, E. H. Chadd, G. F. Liou, K. Bergman, and S. M. Block. Characterization of photodamage to escherichia coli in optical traps. *Biophysical Journal*, 77, (5): 2856–2863, **1999**. 142

[260] S. Suresh, J. Spatz, J. P. Mills, A. Micoulet, M. Dao, C. T. Lim, M. Beil, and T. Seufferlein. Connections between single-cell biomechanics and human disease states: gastrointestinal cancer and malaria. *Acta Biomaterialia*, 1, (1): 15–30, **2005**. 142

[261] X. Yan, R. C. Habbersett, J. M. Cordek, J. P. Nolan, T. M. Yoshida, J. H. Jett, and B. L. Marrone. Development of a Mechanism-Based, DNA Staining Protocol Using SYTOX Orange Nucleic Acid Stain and DNA Fragment Sizing Flow Cytometry. *Analytical Biochemistry*, 286, (1): 138–148, **2000**. 142

[262] S. M. Van Zandycke, O. Simal, S. Gualdoni, and K. A. Smart. Determination of yeast viability using fluorophores. *Journal of the American Society of Brewing Chemists*, 61, (1): 15–22, **2003**. 142

[263] O. Thoumine, P. Kocian, A. Kottelat, and J. J. Meister. Short-term binding of fibroblasts to fibronectin: optical tweezers experiments and probabilistic analysis. *European Biophysics Journal*, 29, (6): 398–408, **2000**. 144

[264] F. Perrin. Mouvement brownien d'un ellipsoide-I. Dispersion dielectrique pour des molecules ellipsoidales. *J. Phys. Radium*, 5: 497–511, **1934**. 144

[265] S. H. Koenig. Brownian motion of an ellipsoid. a correction to perrin's results. *Biopolymers*, 14, (11): 2421–2423, **1975**.

[266] M. X. Fernandes and J. G. de la Torre. Brownian dynamics simulation of rigid particles of arbitrary shape in external fields. *Biophysical Journal*, 83, (6): 3039–3048, **2002**. 144

Appendix

Publications

- C. H. J. Schmitz, K. Uhrig, J. P. Spatz, J. Curtis. Tuning the orbital angular momentum in optical vortex beams. *Optics Express*, 14(15):6604-6612, **2006**.

- C. Mohrdieck, F. Dalmas, E. Arzt, R. Tharmann, M. Claessens, A. Bausch, A. Roth, A. Sackmann, C. Schmitz, J. Curtis, W. Roos, S. Schulz, K. Uhrig, J. Spatz. Biomimetic Models of the Actin Cytoskeleton. *Small*, 3, (6): 1015-1022, **2007**.

- K. Uhrig, R. Kurre, C. Schmitz, J. Curtis, T. Haraszti, A. Clemen, J.P. Spatz. Optical force sensor array in a microSSuidic device based on holographic optical tweezers. *Lab on a Chip*, 9(6):661-668, **2009**.

- T. Haraszti, S. Schulz, K. Uhrig, R. Kurre, W. Roos, C. H. J. Schmitz, J. Curtis, T. Maier, A. Clemen, J.P. Spatz. Biomimetic Models of the Actin Cortex. *Biophys. Rev. Lett.*. *Accepted for publication*.

- S. Hegge*, K. Uhrig*, S. Münter, M. Steinbüchel, R. Kurre, K. Heiss, U. Engel, J. P. Spatz, U. S. Schwarz, K. Matuschewski, F. Frischknecht (*Both authors contributed equally). TRAP-like adhesins and actin coordinate Plasmodium sporozoites adhesion. *In preparation*.

Danksagung

Zu allererst möchte ich mich bei meinem Chef Professor Spatz dafür bedanken, dass er mir dieses anspruchsvolle Thema gegeben hat und mich immer wieder motivierte, am Zug zu bleiben. Dank gilt auch besonders der Aktin Subgruppe in unsererm AK. Hier Dank an Anabel Clemen und Tamas Haraszti für zahlreiche Diskussionen sowie an Anabel noch herzlichen Dank für das sorgfältige Korrekturlesen meiner Arbeit.

Unbedingt bedanken möchte ich mich bei all den Leuten, die aktiv, hands-on an den Experimenten beteiligt waren. Christian Schmitz, der zusammen mit Jennifer Curtis die ursprüngliche optische Pinzette aufgebaut hat und mir alles über die Falle und die Mikrofluidik beigebracht hat, möchte ich für viele Tage im Labor und interessante Gespräche während der Versuche danken.

Rainer Kurre, der als Diplomand bei mir angefangen hat und als Familienvater und Freund nach Norden ausgewandert ist, danke ich für die Zeit und die unzähligen gemeinsamen Stunden an Zippern, Snappern und Netzwerken sowie für die zahlreichen Softwarelösungen, die er kreiert hat.

Timo Maier und Simon Schulz sei gedankt für die Experimente im Pillarteam sowie für die Zusammenarbeit, wenn es um Aktin, Proteinaufbereitung und Reagenzien ging.

Schließlich noch Dank an die nächste Generation von Laseroffizieren. Martin Streichfuss und Friedrich Erbs, die jetzt das Steuer übernehmen werden und mir schon in den letzten Monaten zur großen Hilfe wurden.

Bei den Malaria-Experimenten möchte ich Stephan Hegge, mit dem ich manches Sporozoiten-Grillfest veranstalten durfte, ganz besonders danken. Vielen Dank für diese Kooperation, die langen Abende im Labor und die gesamte Vorarbeit vor jedem

Experiment in der Malaria-Zucht. Hier natürlich auch großen Dank an seinen Chef Freddy Frischknecht für die wissenschaftliche Betreuung unserer Experimente und seine Unterstützung. Weiterhin möchte ich den vielen helfenden Händen danken, die die Experimente erst ermöglicht haben. Thomas Pfeil hat unermüdlich, mit großem Elan Flusszellen gebastelt, Danke hierfür. Christine Mollenhauer hat zusammen mit Michael Bärmann immer für stetigen Proteinnachschub gesorgt. Es war mir eine große Freude, mit ihr zusammenzuarbeiten. Für weitere Aktinlieferungen aus München danke ich ganz herzlich Monika Rusp und der Gruppe von Professor Sackmann.

Natürlich sollen auch unsere weiteren Helfer nicht vergessen werden. Den Sekretärinnen Frau Bozceck, Frau Pfeilmeier und Frau Hess genauso herzlichen Dank wie unserem Faktotum Herrn Richard Morlang, sie alle waren manches Mal sehr hilfreich. Ebenso ein Dank an die Werkstatt, an unsere Elektronik-Experten Herrn Jeschka und Herrn Meinusch sowie selbstverständlich auch an die gesamte Putzbrigade.

Auch an externe Helfer soll hier mein Dank gehen. Cornelia Weber vom AK Fink war immer eine gute Adresse in Aktinfragen und by Myosinbedarf. In Fluoreszenzfragen waren die Mitarbeiter des AK Herten immer hilfreich zur Stelle. Hier besonderen Dank an Kyung-Tae Han für seine Hilfe und die kurzfristige Leihgabe von Filtern und Farbstoffen. Außerdem noch Dank an die Mitarbeiter des AK Tanaka, die oft mit Reagenzien oder Geräten aushelfen konnten. Professor Kierfeld von der TU Dortmund möchte ich danken für seine Hilfe in theoretischen Fragen.

Doch nun zurück zu den zahlreichen Mitgliedern des AK Spatz, die geholfen haben, diese Zeit zu einer besonderen zu machen. Erstmal natürlich Dank an alle Buben aus der Truppe: Daniel Aydin, Babak Hosseini, Theo Lohmüller, der Simme, der Loubeeez und all die anderen, die immer für einen Spaß, einen witzigen Abend oder eine Klassenfahrt zu haben waren, aber auch in allen wissenschaftlichen Belangen oft eine große Hilfe darstellten. Dem ganzen AK Bib möchte ich danken für die spitzen Zeit am Anfang der Doktorarbeit sowie der Frühstücksgemeinschaft Heike, Christian und Babak für den täglichen Morgenstreich. Patrick Heil und Babak Hosseini Dank für Hilfe in softwaretechnischen Fragen und ein extra Dank an Babak und Daniel für ihre Unterstützung bei der Erstellung dieses Manuskriptes. Dank auch an alle anderen Mitglieder des AK Spatz, die immer wieder hilfsbereit zur Seite standen und auf den Ausflügen für eine unvergessliche Atmosphäre gesorgt haben.

Abschließend möchte ich natürlich ganz besonders meiner Familie danken, die mich immer unterstützt und gefördert hat und mein Studium - und damit auch diese Arbeit - erst ermöglich hat. Last but not least schließlich ein großes Dankeschön an meine Freundin Nadine, die mich durch die gesamte Promotion begleitet hat und der ich für jeden einzelnen gemeinsamen Tag danken möchte.

Die VDM Verlagsservicegesellschaft sucht für wissenschaftliche Verlage abgeschlossene und herausragende

Dissertationen, Habilitationen, Diplomarbeiten, Master Theses, Magisterarbeiten usw.

für die kostenlose Publikation als Fachbuch.

Sie verfügen über eine Arbeit, die hohen inhaltlichen und formalen Ansprüchen genügt, und haben Interesse an einer honorarvergüteten Publikation?

Dann senden Sie bitte erste Informationen über sich und Ihre Arbeit per Email an *info@vdm-vsg.de*.

Sie erhalten kurzfristig unser Feedback!

VDM Verlagsservicegesellschaft mbH
Dudweiler Landstr. 99   Telefon +49 681 3720 174
D - 66123 Saarbrücken   Fax   +49 681 3720 1749
www.vdm-vsg.de

Die VDM Verlagsservicegesellschaft mbH vertritt

Printed by Books on Demand GmbH, Norderstedt / Germany